Mind as Metaphor

Mind as Metaphor

A Defence of Mental Fictionalism

ADAM TOON

OXFORD
UNIVERSITY PRESS

Great Clarendon Street, Oxford, OX2 6DP,
United Kingdom

Oxford University Press is a department of the University of Oxford.
It furthers the University's objective of excellence in research, scholarship,
and education by publishing worldwide. Oxford is a registered trade mark of
Oxford University Press in the UK and in certain other countries

© Adam Toon 2023

The moral rights of the author have been asserted

All rights reserved. No part of this publication may be reproduced, stored in
a retrieval system, or transmitted, in any form or by any means, without the
prior permission in writing of Oxford University Press, or as expressly permitted
by law, by licence or under terms agreed with the appropriate reprographics
rights organization. Enquiries concerning reproduction outside the scope of the
above should be sent to the Rights Department, Oxford University Press, at the
address above

You must not circulate this work in any other form
and you must impose this same condition on any acquirer

Published in the United States of America by Oxford University Press
198 Madison Avenue, New York, NY 10016, United States of America

British Library Cataloguing in Publication Data

Data available

Library of Congress Control Number: 2023930048

ISBN 978–0–19–887962–6

DOI: 10.1093/oso/9780198879626.001.0001

Printed and bound in the UK by
Clays Ltd, Elcograf S.p.A.

Links to third party websites are provided by Oxford in good faith and
for information only. Oxford disclaims any responsibility for the materials
contained in any third party website referenced in this work.

For Laura and Natalie

'But must I needs want solidness, because
By metaphors I speak?'

(John Bunyan, *The Pilgrim's Progress*)

Contents

Preface	xi
Acknowledgements	xv
Introduction: The Story of the Ghost in the Machine	1
1. Making Up Minds	9
1.1 Folk Psychology	9
1.2 Mind as Machinery	11
1.2.1 The Representational Theory of Mind	11
1.2.2 Eliminativism	14
1.3 Mind as Metaphor	15
1.3.1 Metaphor and Make-believe	15
1.3.2 Myths and Metaphors	18
1.4 Implications	24
1.4.1 Making Sense of the Folk	24
1.4.2 Does the Mind Exist?	30
1.5 Conclusion	32
2. Contrasts, Causes, and Consciousness	33
2.1 Anti-Cartesianism	33
2.1.1 Behaviourism	33
2.1.2 Ryle	36
2.1.3 Instrumentalism	38
2.1.4 Dennett	41
2.1.5 Prefix-fictionalism	46
2.2 Explaining Behaviour	48
2.2.1 Wires and Pulleys	48
2.2.2 Imagined Mechanisms	51
2.3 Self-knowledge	52
2.3.1 Finding Out about Ourselves	53
2.3.2 Talking to Ourselves	56
2.4 Conclusion	59
3. Minds and Representations	60
3.1 Intentionality	60
3.2 Mental Representations	61
3.3 Fictionalism and Intentionality	63
3.3.1 Thoughts	63
3.3.2 Language	65
3.3.3 Taking Stock	68

X   CONTENTS

3.4 Cognitive Collapse	68
3.4.1 The Problem	69
3.4.2 Pretending	71
3.5 Why Not Representationalism?	75
3.5.1 Fictions and Hypotheses	76
3.5.2 Mental Representations as Real Fictions	79
3.6 Conclusion	83
4. Minds and Materials	84
4.1 Material Culture	84
4.1.1 Thoughts and Things	84
4.1.2 The Extended Mind Thesis	85
4.1.3 ExM and Representationalism	87
4.2 Minds, Materials, and Metaphors	87
4.2.1 Materials as Metaphors	88
4.2.2 Materials as Minds	89
4.3 Defending ExM	91
4.3.1 Common Sense	92
4.3.2 Intentionality	93
4.3.3 Cognitive Science	95
4.3.4 Drawing Boundaries	96
4.4 Folk Miracles?	99
4.5 Conclusion	104
5. Inquiring Minds	105
5.1 Minds and Inquiry	105
5.2 Epistemology as Fiction	108
5.2.1 Concepts	109
5.2.2 Knowledge	115
5.2.3 Understanding	119
5.3 Reason, Tools, and History	122
5.4 Conclusion	126
Epilogue: An Uneasy Presence . . .	128
References	131
Index	139

Preface

Now that it's finished, I realize that I wrote this book backwards. My work first began with a topic that now makes its appearance only in the book's final chapter, namely the importance of material culture in inquiry. I was originally trained in philosophy of science, and have long been fascinated by historical and sociological work on the role of material culture in scientists' reasoning, from formulas and diagrams to field guides and three-dimensional, physical models. If philosophers wanted to understand scientific reasoning, it seemed to me, we should focus on what scientists do with these bits of paper, wood, or plastic—not speculate about what might be going on inside their heads. I wrote my PhD on scientific models, which become the subject of my first book (*Models as Make-Believe*, Palgrave Macmillan, 2012). In the book, I argued that we can understand scientific models by looking to children's games of make-believe. One of my examples was the ball-and-stick models of molecules that many of us remember from school chemistry classes. Just as a child imagines a doll to be a baby so too, I argued, scientists imagine a ball-and-stick model to be a molecule.

Towards the end of my PhD, I happened to see Andy Clark give a lecture in Cambridge on the extended mind thesis. The central claim of the extended mind thesis is that thinking doesn't always happen inside the head. I was immediately struck by the parallels between this idea and the historical and sociological literature I knew on the importance of material culture in science. A few years later, I was fortunate enough to be awarded a Marie Curie Fellowship to explore the implications of the extended mind thesis for philosophy of science. As part of that project, I wrote papers on the consequences of the extended mind thesis for a range of issues in philosophy of science, such as empiricism and the use of scientific instruments ('Empiricism for cyborgs', *Philosophical Issues*, 2014) and the nature of scientific understanding ('Where is the understanding?', *Synthese*, 2015). As time went on, however, I began to harbour misgivings. Although I was convinced that the extended mind thesis captured an important insight about the role of material culture in thinking, I worried that it retained too much of the traditional picture of the mind as an inner world of representations.

It was around this time that I had another piece of good fortune, and was invited to give a talk on mental fictionalism at a workshop at the University of Edinburgh, organised by Adrian Downey, Joe Morrison, and Mark Sprevak. As I wrote my talk, I began to realize that the ideas I had explored in my earlier work on models might offer a way to resist the traditional picture of the mind, while also explaining its hold over us. Perhaps we ordinary folk pretend that the mind is an

xii PREFACE

inner world, in the same way that scientists pretend that falling objects are point masses or that surfaces are free from friction? Flushed with enthusiasm, I wrote an initial defence of this idea ('Fictionalism and the folk', *The Monist*, 2016). At roughly the same time, I had several helpful and enjoyable discussions with John Dupré at Exeter, about the extended mind thesis and its similarity to themes in Ryle and Wittgenstein. When I began to teach at Exeter, I decided to offer a course on Ryle's *The Concept of Mind* (1949). Engaging with Ryle's ideas with my students was a joy, and encouraged me to think that a fictionalist approach was worth developing further—even if I doubted whether Ryle himself would approve.

Since then, I have continued to develop a fictionalist approach to the mind as best I can. This has meant figuring out how it might avoid some seemingly insurmountable problems and exploring its implications for key issues in the philosophy of mind (and sometimes wondering if I had bitten off more than I could chew). Eventually, I returned to the topic that got me hooked in the first place, and tried to show how a fictionalist approach to the mind could do justice to the importance of material culture in inquiry. As I see it, fictionalism allows us to avoid the trap of thinking that our interactions with material culture are merely a pale reflection of what is going on inside our heads. At the same time, it explains why the extended mind thesis does not go far enough. It is not simply that the mind sometimes extends into the world. Instead, our entire conception of the 'inner world' of the mind is a metaphorical projection of the 'outer world' of material culture. The upshot is that the history of material culture—including our scientific culture—is the history of the mind.

Many people have helped me while I've been writing this book. I would like to thank all my colleagues at the University of Exeter for making it such a stimulating and enjoyable place to work. Thanks especially to Giovanna Colombetti, Adrian Currie, Jonathan Davies, John Dupré, Shane Glackin, Stephan Guttinger, Iza Kavedzija, Joel Krueger, Staffan Müller-Wille, Celso Neto, Andy Pickering, Nigel Pleasants, Anna Rabinovich, Tom Roberts, Richard Seaford, Darren Schreiber, Ed Skidelsky, Kirsten Walsh, and Sam Wilkinson. All of them have patiently listened to me rattle on about mental fictionalism at one time or another, and have offered valuable comments and criticism. A special thank you also to Sabina Leonelli, for all her kind support and encouragement with the project. I have presented ideas from this book in many seminars and reading groups at Exeter over the years, including the Exeter Research Exchange and the Cognition and Culture reading group. My thanks to the audiences at these events for helpful feedback. I would also like to thank the undergraduate students who took my course on *The Concept of Mind* for some hugely enjoyable discussions of Ryle's views, as well as the Master's students who took my classes on mental fictionalism for a module on Current Issues in Mind and Cognition. Finally, thank you to my PhD students, especially Antonis Antoniou, Kane Baker, Ric Sims, and Necmiye Yuksel, for lots of enjoyable conversations and debates over the years.

PREFACE xiii

Further afield, I am especially grateful to Tamás Demeter and Ted Parent, with whom I organized a conference on mental fictionalism in Budapest in October 2019, and edited a volume on the same topic (*Mental Fictionalism*, Routledge, 2022). Thanks also to those who participated in the conference and contributed such excellent papers for the volume, including Tim Crane, Julianne Chung, Greg Currie, Dan Dennett, Adrian Downey, Zoe Drayson, Kati Farkas, Dan Hutto, László Kocsis, Bill Lycan, Miklós Márton, Bruno Mölder, Krisztián Pete, Amber Ross, Julia Tanney, János Tőzsér, Meg Wallace, and Sam Wilkinson. I am also grateful to Dan O'Brien, Mark Cain, Andy Clark, Matteo Colombo, Steven French, Stacie Friend, Roman Frigg, Liz Irvine, Max Jones, Martin Kusch, Jeff Kochan, Mary Morgan, Robert Meunier, James Nguyen, Fiora Salis, Mark Sprevak, Mog Stapleton, Martin Thomson-Jones, and Frédéric Vallée-Tourangeau, who have all been kind enough to offer feedback on ideas developed in this book. And, of course, a special thank you to Kendall Walton, whose ideas have been such an important source of inspiration, both for my earlier work on models and for the approach to the mind that I develop in this book. Finally, I would like to thank Peter Momtchiloff at Oxford University Press, as well as two anonymous readers, who provided extremely helpful and constructive comments on an earlier version of the manuscript.

I have presented ideas from this book at various seminars, workshops, and conferences over the years, including the 'Mental Fictionalism' conference in Edinburgh in July 2014, the 'Shaping the Trading Zone' workshop in Leeds in September 2014, a departmental seminar at the University of Bristol in May 2015, the 'Scientific Fictionalism' conference in London in June 2015, the Biennial Conference for Philosophy of Science in Practice in Aarhus in June 2015, the 'Extended Mind' workshop in Cardiff in April 2016, the Annual Conference of the British Society for the Philosophy of Science in Cardiff in July 2016, a departmental seminar at Queen's University Belfast in November 2016, a departmental seminar at the University of Edinburgh in December 2016, the 'Integrated History and Philosophy of Science' workshop in Nottingham in April 2017, the 'Metaphors of the Mind' workshop at the Swiss Institute in Rome in May 2017, a departmental seminar at the University of Bielefeld in June 2017, the 20th Conference of the Society for Philosophy and Technology in Darmstadt in June 2017, the 'Reconceiving Cognition' conference in Antwerp in June 2018, a departmental seminar at Sun Yat-sen University, Guangzhou in March 2018, a departmental seminar at Oberlin College in April 2018, a Royal Institute of Philosophy public lecture at Blackwell's Bookshop in Oxford in November 2018, the 'Andy Clark and his Critics' conference in Edinburgh in May 2019, a Logic, Epistemology and Metaphysics Seminar at the Institute of Philosophy in March 2019, the 93rd Joint Session of the Aristotelian Society and the Mind Association in Durham in July 2019, the 'Mental Fictionalism' conference in Budapest in October 2019, a departmental seminar at University College London in January

xiv PREFACE

2020, and the 'Science and Philosophy of Imagination' conference in Bristol in September 2021. I would like to thank the organizers and audiences at all these events for helpful comments and questions.

Finally, I would like to thank my family for all their love and support, especially my wife, Angela, and our daughters, Laura and Natalie. I gave my first talk on mental fictionalism a few months before our eldest daughter, Laura, was born. In the time it's taken me to get my ideas straight and write this book, both her and her sister, Natalie, have learned how to crawl, walk, run, talk, count, read, sing, dance, swim, ride a bike, and roller-skate. It all rather puts the pace of academic work to shame. Still, I do manage to keep up sometimes, and sharing in their adventures is the greatest joy and privilege of my life.

Acknowledgements

Parts of this book incorporate revised material from the following articles. I am grateful to the publishers for permission to use that material here.

The Story of the Ghost in the Machine. In S. Sedivy (ed.), *Art, Representation and Make-Believe: Essays on the Philosophy of Kendall L. Walton*. Abingdon, Oxfordshire: Routledge. © Taylor & Francis. Reproduced with permission of The Licensor through PLSclear.

Fictionalism and the folk. *The Monist*, 99: 280–95. Reproduced with permission of Oxford University Press.

Fictionalism and intentionality. In A. Toon, T. Demeter, and T. Parent (eds), *Mental Fictionalism: Philosophical Explorations*, Abingdon, Oxfordshire: Routledge. © Taylor & Francis. Reproduced with permission of The Licensor through PLSclear.

Minds, materials and metaphors. *Philosophy*, 96: 181–203. © The Royal Institute of Philosophy, 2020. Reprinted with permission of Cambridge University Press.

Introduction

The Story of the Ghost in the Machine

Once upon a time, many philosophers might have been happy to think of the mind as a special substance, distinct from the body—a spirit or soul, perhaps, or mind-stuff. Lurking in a hidden and private realm that each of us possessed, this strange and wondrous substance was the repository of our beliefs, desires, hopes, and fears. Although officially not part of space, it was tempting to think of it as our 'inner' world, in contrast to the 'outer' world of tables, chairs, and mountains. When our bodies moved around the outer world (placing us in front of a sunset, for example, or having us bite into a rotten apple) changes resulted in our inner world (a belief in the beauty of nature, perhaps, or the desire to pick a different apple). In turn, changes within our inner world might bring about changes to our bodies (we might linger a while in front of the view or throw the apple away in disgust). Despite these frequent interactions, however, the inner world of the mind was quite unlike the material stuff that makes up tables and chairs—or, indeed, the human body. And, of course, unlike our bodies, it might also hope to live happily ever after.

Nowadays, most philosophers find the existence of any kind of non-physical substance a bit too spooky to believe in. Instead, most will locate the mind in a particular part of the material world (albeit an especially marvellous and complex part) namely, the human brain. And yet for many authors within philosophy, cognitive science, and elsewhere, the basic structure of the traditional picture of the mind remains intact. The mind is still taken to be a kind of private, inner world that houses our mental states—our beliefs, desires, hopes, fears, and the rest. Indeed, it is now quite literally an inner world: rather than existing somehow outside space, it is safely locked away inside our skulls. Just as before, the state of the outer world and our bodies can affect this inner world (the sunset can still induce a belief in the beauty of nature, or the apple a desire for a different piece of fruit) and, in turn, this inner world can direct the movement of our bodies (our belief can still cause us to linger a while, and our desire to throw the apple away). Of course, since our inner world is now resolutely material, these interactions seem rather less mysterious than before, and our hopes for immortality are somewhat diminished. And yet, despite all this, the basic picture of the mind as an inner world is retained.

Mind as Metaphor: A Defence of Mental Fictionalism. Adam Toon, Oxford University Press. © Adam Toon 2023.
DOI: 10.1093/oso/9780198879626.003.0001

2 MIND AS METAPHOR

Not all have accepted this vision of the mind, of course. Gilbert Ryle's *The Concept of Mind* is a wonderfully irreverent assault upon the traditional picture of the mind (or 'official doctrine'), which Ryle memorably dubbed 'the dogma of the ghost in the machine' (1949, 17). According to Ryle, the view is absurd. It cannot make sense of our ordinary talk about the mind, and it renders even our humdrum, everyday knowledge of the thoughts and feelings of others (and ourselves for that matter) utterly mysterious. Ryle's critique is normally thought merely to tell against *dualism*—the view that the mind and body are distinct substances. And yet many of his arguments can also be levelled against more recent, materialist versions of the approach. For Ryle's main complaint against the traditional picture is not that it takes the mind to be non-physical, but that it takes it to be a kind of hidden mechanism—a set of unseen 'wires and pulleys' directing us from behind the scenes, locked away in a 'secret grotto' (Ryle 1949, 114; see also Tanney 2009). It makes little difference whether we take this hidden mechanism to be made of mysterious mind-stuff or wonderfully intricate grey matter.

For many authors, the official doctrine simply expresses our ordinary, common-sense picture of the mind. For Ryle, by contrast, it is a 'philosopher's myth' (1949, 17). It is a category mistake that philosophers have inflicted upon our ordinary language for talking about the mind. According to Ryle, the chief culprit is Descartes. Indeed, the 'inner grotto' view of the mind is often referred to simply as *Cartesianism* (and its more recent, materialist variant as *neo-Cartesianism*). Some will question whether Ryle's image of the ghost in the machine captures the subtleties of Descartes' view, of course. Others have questioned Ryle's claim that Descartes is the source of this picture of the mind. In an early review of *The Concept of Mind*, Stuart Hampshire objected to Ryle's account of the ghost's origins:

> so far from being imposed on the plain man by philosophical theorists, and even less by seventeenth century theorists, the myth of the mind as a ghost within the body is one of the most primitive and natural of all the innumerable myths which are deeply imbedded in the vocabulary and structure of our languages.... Professor Ryle throughout represents philosophers as corrupting the literal innocence of common sense speech with alien metaphors; in this he not only greatly exaggerates the influence of philosophers, and particularly of Descartes, on the forms of common speech, but (more seriously) neglects the fact, or rather the necessity, that the forms of common speech and its modes of description should be permeated with such metaphors, most of which can ultimately be traced back to underlying myths and imaginative pictures.
>
> (1950, 237; emphasis in original)

In a sense, this book can be seen as an attempt to follow up on Hampshire's insight. The notion of the mind as an inner world, I shall suggest, is not merely a philosopher's myth. Instead, it is a metaphor (or, rather, a set of metaphors) that lie at the heart of our ordinary talk about the mind. Unlike Hampshire, I will not claim that the metaphors I consider are primitive or natural, or that they would be familiar to the Greeks or Romans. My aim is not to answer the historical question of when these metaphors became part of our concept of mind. My aim is simply to show that they *are* part of that concept, and that this is important for our understanding of the mind. We talk about people *as if* they have mysterious inner worlds that house their beliefs and desires. And yet, I shall argue, we do not—and should not—take such talk too seriously. To put the point slightly differently, the idea of the mind as an inner world is not a philosopher's myth, but a folk story. It is a story that we tell to help make sense of people and the way they behave. Like Ryle, I think that the story is not true. Indeed, I think it cannot be true. Like much science fiction or fantasy, when we examine its plot too closely we find that it stretches our credulity and even harbours latent contradictions. And yet it is not merely a story told by philosophers and foisted upon the unsuspecting folk. Instead, the story of the mind as an inner world is crucial to our ordinary talk about people and their mental states. Indeed, it is a story that we cannot avoid telling.

This book will therefore steer a course that lies somewhere between the official doctrine and Ryle's view. The official doctrine is correct to say that we talk about the mind as an inner world. It is not simply that philosophers have entirely misread our ordinary language. The trouble arises when they start to take this talk literally, rather than metaphorically. Put simply, if Ryle's view neglects the importance of our metaphors, the official doctrine forgets that they are just that— metaphors. Of course, Hampshire is not the only author to have noted the role of metaphors in our descriptions of the mind. Julian Jaynes (1976) and Daniel Dennett (1991a) have both written about the importance of metaphor for under-standing our conscious experience. Douwe Draaisma (2000) has traced the history of different metaphors found in theories of memory, from wax tablets to holo-grams. Especially close to our interests here, Derek Melser (2004) has emphasized the importance of metaphor in our ordinary talk about the mind. Melser argues, as I do, that the notion of the mind as an inner world arises from taking our metaphors too literally. His discussion focuses especially on the noun 'mind', which he claims to be a nominalization of the old-fashioned verb 'to mind'. Like most philosophers, however, Melser takes a fairly dim view of these metaphors: he argues that they are a 'mess' and should be put aside. In contrast, I will emphasize the positive aspects of our metaphorical talk about the mind. As I see it, these metaphors are not merely peripheral, nor are they dispensable. Instead, they lie at the heart of our ordinary conception of the mind (or *folk psychology*).

My discussion will focus on what we might loosely call *thought,* understood broadly to include *propositional attitudes* (e.g. beliefs, desires, and intentions), as well as related aspects of the mind (e.g. concepts, reasoning, and understanding).

4   MIND AS METAPHOR

In this area, the traditional view tends to take a particular form, known as the *representational theory of mind* (or *representationalism*). The core idea of the representational theory of mind is that thoughts are inner representations (or *mental representations*). For example, having a belief involves having a mental representation with the right content inside your head that causes you to behave in a particular way, while a desire might involve a mental representation with the same content that causes different sorts of behaviour. Reasoning involves having a succession of mental representations that follow each other in the right manner. Possessing a concept involves having a different sort of mental representation that allows you to categorize objects in your experience. And so on. The representational theory of mind has a long and distinguished history. Versions of it can be found in the work of Hobbes, Descartes, Locke, Berkeley, Hume, and many others. It also remains the dominant conception of the mind today. Indeed, the representational theory of mind forms a central part of the foundations of cognitive science, the interdisciplinary study of the mind that was born in the 1950s and '60s.

My own view will stand in sharp contrast to the representational theory of mind. I will argue that mental representations are useful fictions. People do not really have representations inside their heads, but talking *as if* they do helps us to make sense of their behaviour. There are external, public representations, of course—spoken and written language, pictures, maps, diagrams, and the rest. When we talk about thoughts, I will argue, we project this *outer* world of material culture onto a metaphorical *inner* world of the mind. Of course, some of our talk about thoughts is overtly metaphorical or figurative. Few will doubt that we are speaking figuratively if we say that we are making a mental note or putting an unpleasant thought to the back of our mind. My claim will be more controversial. I will argue that metaphor underpins even the most apparently sober and straightforward things we say about people's thoughts, like 'Ruth wants to buy some bread' or 'Mark believes the No. 73 bus goes to Oxford Street.' Explaining and predicting people's behaviour by attributing beliefs and desires to them is commonly taken to lie at the heart of folk psychology. According to the representational theory of mind, this is a practice of folk theorizing: we are making hypotheses about people's inner machinery to explain and predict what they will do. In contrast, I will argue that we are speaking metaphorically. Our metaphors greatly enrich our language, and we would be lost without them. But we must not forget that the inner world that they conjure up is only a fiction.

Since I focus on thought, I will have relatively little to say about other aspects of the mind, such as sensations or emotions. We use metaphors here too, of course. Once again, some of these metaphors are fairly obvious. We say that we've got butterflies in our stomach or that we're feeling blue. Others are more subtle, and perhaps more problematic. For instance, Ryle criticizes proponents of the official doctrine for talking about sensations as if they were strange, private objects that only their owner can observe (1949, chapter VII). By talking in this way, he thinks, we encourage our temptation to think of the mind as a hidden world, in which such objects can be housed. Although I am sympathetic to Ryle's views on the nature of sensations, I will not try to defend them here. For the most part, I will simply acknowledge that we have sensations, without defending any particular position on how they are to be understood. For our purposes, the important point will be that acknowledging these aspects of our mental lives need not return us to the traditional view of the mind. In particular, it need not return us to the idea that our thoughts exist within some private, inner realm, locked away from the outside world.

The view I shall defend is called *mental fictionalism*. Fictionalism is now a popular position in many fields, including metaphysics, moral philosophy, philosophy of language, and philosophy of mathematics. The core idea of fictionalist approaches is that talk in some area can be useful even if it is not straightforwardly true. For example, a realist about mathematics might argue that mathematical claims like '2 + 2 = 4' are true if they correctly describe a timeless, Platonic realm of mathematical entities, such as numbers or sets. By contrast, a fictionalist about mathematics might argue that this Platonic realm does not exist and therefore that, strictly speaking, mathematical claims are false. And yet, the fictionalist insists, mathematical claims can still be useful. For example, they can help us to talk about ordinary physical objects. In fact, the fictionalist might go further and suggest that, once we understand mathematical claims correctly, we see that they can express truths after all—it's just that these truths are not about a mysterious world of mathematical entities, but about everyday things like tables and chairs.

Mental fictionalism—or, at least, the version of mental fictionalism that I shall defend in this book—has a similar structure. The core idea of a fictionalist approach to the mind is that talk about mental states can be useful even if it is not straightforwardly true. The traditional view holds that our ordinary claims about the mind like 'Mark believes the No. 73 bus goes to Oxford Street' are true if they correctly describe a mysterious collection of mental entities, such as beliefs and desires, hidden away in a secret, inner realm. By contrast, mental fictionalism claims there is no such inner realm—either physical or non-physical. Taken literally, our talk about it is false. And yet, the mental fictionalist insists, our talk about this inner realm can still be useful. For example, it can help us to talk about Mark's behaviour. In fact, the mental fictionalist can go further and suggest that, once we understand talk about the mind correctly, we see that it can express truths after all—it's just that these truths are not about a mysterious world of mental entities, but about the everyday things that people say and do, like saying they plan to go to Oxford Street or jumping on a bus.

To develop this approach, I will draw on an influential theory of fiction due to the philosopher of art Kendall Walton (1990). At the heart of Walton's approach is the idea that works of fiction can be understood as objects used in games of

6 MIND AS METAPHOR

make-believe, rather like dolls and hobbyhorses. Walton (1993) also applies this framework to give an analysis of metaphor and figurative language that has been influential amongst fictionalists in many domains (e.g. Crimmins 1998; Kroon 2001; Yablo 2001; Joyce 2005). In my own work, I have drawn on Walton's ideas to develop a theory of scientific models (Toon 2010, 2011, 2012). It is characteristic of scientific models, I have argued, that they represent the world by asking us to imagine it as other than it is—a surface is imagined to be free of friction, the earth is imagined to be a perfect sphere, economic agents are imagined to be ideally rational, and so on. In a similar vein, I will suggest, much of our ordinary talk about the mind involves imagining people as other than they are: we imagine them to have inner worlds that harbour their beliefs and desires. As we shall see, however, applying this idea to our talk about the *mind*—rather than talk about planets or inclined planes—is far from straightforward.

Despite the prominence of fictionalism in other areas of philosophy, mental fictionalism has found few supporters (exceptions include Demeter 2013; Toon 2016; Wallace 2016, 2022). One reason for this might be that mental fictionalism faces particular difficulties that do not confront other forms of fictionalism. Perhaps the most serious such difficulty is that mental fictionalism seems flatly incoherent. The mental fictionalist tells us to treat talk about the mind as a useful fiction. And yet doesn't treating something as a fiction require the existence of certain mental states (or acts or processes)? For instance, a moment ago I said that we *imagine* that people have inner worlds—and yet surely imagination is a mental state (or act or process) if anything is. I will try to show how mental fictionalism can avoid this worry, as well as other important objections that have been levelled against it. I will also try to demonstrate some of the attractions of mental fictionalism. Doing so will require us to explore its implications for a wide range of issues, such as the nature of intentionality and the relationship between mind and material culture. It will also require us to rethink central aspects of the fictionalist framework itself, such as its use of notions like imagination and make-believe.

In recent years, the traditional foundations of cognitive science have come under increasing threat from a range of new movements within the field. These movements are often grouped under the heading of '4E cognition', since they see cognition as variously *embodied*, *embedded*, *extended*, and *enacted* (for helpful overviews, see Robbins and Aydede 2009 and Menary 2010). The central theme of much of this work is that cognition should be seen not only (or, perhaps, not at all) as a matter of the manipulation of internal representations, but as a phenomenon that involves the interplay of brain, body, and environment. For example, many studies emphasize that cognition includes interaction with material culture, such as notebooks, maps, or diagrams. The different movements are by no means equivalent, however. They differ in their central theoretical commitments and the degree to which they reject the assumptions of traditional cognitive science.

For example, defenders of the *extended mind thesis* retain the idea that cognition should be understood as the manipulation of mental representations—they simply claim that these representations can sometimes extend outside the head (Clark and Chalmers 1998). By contrast, some advocates of enactive approaches claim that cognition can often be explained without appealing to mental representations at all (Hutto and Myin 2013).

My hope is that mental fictionalism can contribute to these ongoing debates within cognitive science. Unlike the extended mind thesis, mental fictionalism rejects the representational theory of mind. According to the fictionalist, mental representations do not exist—either inside or outside the head. And yet, unlike advocates of enactive approaches, mental fictionalism need not urge cognitive science to stop talking about mental representations. According to the fictionalist, science is full of useful fictions, from ideal oscillators to ideally rational agents. Even if such things do not exist, they can still do useful work for us. In this way, fictionalism might offer a new stance within recent debates over mental representation and its proper place within cognitive science. At the same time, I shall also suggest that mental fictionalism gives us a better understanding of the relationship between mind and material culture. It is not simply that we sometimes use notebooks, maps, and the rest in order to help us think and reason. Instead, the world of material culture also shapes our vision of the mind as an inner realm.

The discussion will proceed as follows:

In Chapter 1 ('Making up Minds'), I will begin by considering the nature of our ordinary talk about the mind (or *folk psychology*). I will also consider the most common interpretation of folk psychology, namely that it should be understood as a theory of the mind as an inner realm. This suggests two alternatives: either our folk theory is true or it is false. The first option is taken by the *representational theory of mind* and the second by *eliminativism*. After briefly considering each of these positions, I will introduce mental fictionalism and show how it offers a new way of understanding folk psychology and the nature of mental states.

In Chapter 2 ('Contrasts, Causes, and Consciousness'), I will further develop and motivate mental fictionalism by comparing it to a number of well-known alternative positions, such as *behaviourism* and *instrumentalism*. I also consider how fictionalism relates to Ryle's view of the mind and Daniel Dennett's influential work on the *intentional stance*. I shall try to show that fictionalism has advantages over each of these positions. I will also respond to two important objections that might be levelled against mental fictionalism: that it overlooks the fact that mental states are causes, and that it cannot explain the immediate, conscious awareness that we are supposed to have of our own mental states.

In Chapter 3 ('Minds and Representations'), I will explore the implications of mental fictionalism for the nature of *intentionality*. An important feature of mental states is that they can represent things in the world. According to the representational theory of mind, the intentionality of thought is to be explained in

8 MIND AS METAPHOR

terms of the content of mental representations. Since fictionalism denies the existence of these representations, it must adopt a different approach to intentionality. Once we understand the nature of its approach to intentionality, I shall argue, we can see that mental fictionalism can avoid the charge of incoherence (or *cognitive collapse*). I will also explain why I think we should prefer fictionalism to realism about mental representations.

In Chapter 4 ('Minds and Materials'), I will show that mental fictionalism offers a new perspective on the relationship between mind and material culture. According to the extended mind thesis, items of material culture can form part of the material basis for our mental states. As we have noted already, the extended mind thesis assumes a representationalist account of mental states. I will argue that, by rejecting representationalism and turning to fictionalism, we can retain the important insights underlying the extended mind thesis, while avoiding its more problematic consequences. I will close the chapter by responding to the worry that fictionalism makes the success of folk psychology a miracle.

In Chapter 5 ('Inquiring Minds'), I will develop a fictionalist approach to the nature of knowledge and inquiry. Many of our greatest thinkers about knowledge—such as Descartes, Locke, Berkeley, and Hume—have accepted the basic vision of the mind as an inner world of representations. Their disagreements concern the inhabitants of this inner world and its relationship to the world outside—if, indeed, there *is* a world outside the mind. I will sketch an alternative, fictionalist approach to the nature of concepts, knowledge, and understanding. An important implication of this approach is that the mind has a *history*. From mathematical diagrams in ancient Greece to the invention of computer simulations, new tools and practices have dramatically transformed the nature of the human mind and its limits.

1

Making Up Minds

What are mental states? When we say that someone has a particular belief or desire, are we making a claim about what is going on inside their head? If we are, might our future science of the mind show that we are wrong? Might it turn out that mental states do not exist? This chapter will introduce mental fictionalism and show that it offers a new approach to these longstanding questions about the mind. I will begin by introducing our ordinary talk about the mind or *folk psychology* (Section 1.1) and the idea that it is best understood as a theory of the mind as an inner realm (Section 1.2). I will then introduce mental fictionalism and show that it offers a new interpretation of ordinary talk about the mind (Section 1.3). Finally, I will explore some of the implications of this approach for our understanding of folk psychology and the nature of mental states (Section 1.4).

1.1 Folk Psychology

Our everyday language is full of talk about the mind. Or, to put it more precisely, our everyday language is full of talk about people's beliefs, desires, decisions, reasoning, hopes, fears, knowledge, and understanding...as well as countless other mental states, acts, and processes. We attribute such states (or acts or processes) to people all the time, often without stopping to think much about what we are doing. We also use them to explain and predict people's behaviour. If you walked into the kitchen and saw me reaching for the tea caddy, you would likely attribute a whole host of mental states to me with relatively little effort— even if you had never met me before and knew nothing about my insatiable tea habit. For a start, you would almost certainly assume that I *want* a cup of tea and *believe* that there are teabags in the caddy. You would probably also assume that I *want* a mug and some boiling water too, that I *believe* that tea isn't poisonous, that I *know* that tea comes in teabags, that I *understand* how to make a cup of tea, and so on. Attributing these (and many other) mental states to me allows you to *explain* why I'm reaching for the tea caddy. If you know the tea caddy is empty, it also allows you to *predict* that I'll be disappointed when I look inside it—and perhaps that I'll head off to the shops to buy some tea, or have a cup of coffee instead.

How are we to make sense of this practice? What exactly are we doing when we attribute mental states to people—like wanting a cup of tea or believing that there

Mind as Metaphor: A Defence of Mental Fictionalism. Adam Toon, Oxford University Press. © Adam Toon 2023.
DOI: 10.1093/oso/9780198879626.003.0002

10   MIND AS METAPHOR

are teabags in the tea caddy? What do such attributions say about people? What makes them true or false? In short, how should we understand ordinary talk about the mind? Our answers to these questions will influence our approach to other issues too, like the question of what mental states *are*. It need not determine our answer to this question, of course. Our ordinary talk might turn out to be misleading in certain ways. Indeed, there is a sense in which this is precisely what I will argue in this book. Still, our view about the nature of mental states cannot ignore the way that we talk about them. For this reason, our exploration of mental fictionalism will begin with our ordinary talk about the mind. To simplify matters a little, I will focus on our talk about what we might call *thoughts*— understood broadly to include *propositional attitudes* (e.g. beliefs, desires, and intentions), as well as related aspects of the mind (e.g. concepts, reasoning, and understanding). How should we understand our talk about people's thoughts? What does this show us about the nature of thought?

Three features of our talk about thoughts are worth noting at the outset.

The first is that we seem to take thoughts to be *about* things in the world. Put differently, thoughts *represent* the world. They have *contents*. For example, my belief that there is tea in the tea caddy is *about* tea and the tea caddy. It *represents* the tea caddy as having tea in it. Its *content* is that there is tea in the tea caddy. In philosophers' terminology, we can say that beliefs, desires, and other mental states have *intentionality*.

The second is that our talk about thoughts seems to involve *causal* claims. We talk as if our thoughts can cause, and can be caused by, events in the world. For example, a long meeting might cause my desire for a cup of tea which, in turn, causes me to reach for the kettle. We also talk as if thoughts can cause other thoughts. For example, my belief that you can buy some teabags at a corner shop, together with my belief that there is a corner shop on Union Road, causes my belief that I can buy some teabags on Union Road.

The third feature of our ordinary conception of thoughts is more difficult to express. It is that we seem to take thoughts to be somehow 'hidden'. At least, we talk as if we cannot see (or hear) other people's thoughts in the same way that we can see (or hear) what they are currently doing or saying. We must be careful here. It is not that people's thoughts are always unknown to us. If you walk into the kitchen and I say, 'we've run out of tea', you might know perfectly well that I believe that we've run out of tea. Still, it seems wrong to say that my belief just *is* my saying this to you. Instead, we might be more inclined to say that my belief *caused* me to say it. At any rate, we are also well aware that people do not always say what they believe, or believe what they say.

Each of these features of our ordinary talk about thoughts will be interpreted differently by different theories of the mind. As we have noted already, some authors (including me) will argue that certain aspects of our ordinary talk are misleading. For instance, some will deny that we make causal claims in the way that I have suggested, or dispute the idea that we take thoughts to be in some sense hidden. We will consider each of these issues as we go on. For the moment, our aim is simply to identify some key aspects of our ordinary talk about thoughts, which any theory must explain (or, perhaps, explain away). Following many authors, I will call our ordinary talk about the mind *folk psychology*. The term 'folk psychology' can be misleading. I shall use it in a fairly neutral way to refer to our ordinary talk about the mind—whatever form such talk might happen to take. Often, though, the term 'folk psychology' is taken to imply a particular *interpretation* of our ordinary talk about the mind—namely, that it constitutes a kind of rudimentary scientific theory of the mind as an inner realm. We can now turn to examine this idea.

1.2 Mind as Machinery

One common way of understanding our ordinary talk about the mind is to see it as an attempt to describe what goes on inside our heads. In this view, folk psychology is a theory of our inner machinery. For want of a better term (and with no pretence to historical accuracy), let us call this approach to folk psychology *Cartesianism*. If we accept the Cartesian approach, then it seems there are, broadly speaking, two possibilities: either folk psychology is a *true* theory of our inner machinery or it is *false*. The first option is taken by the *representational theory of mind* and the second by *eliminativism*. Let us consider both positions in turn.

1.2.1 The Representational Theory of Mind

The core idea of the *representational theory of mind* (or *representationalism*) is that our mental life should be understood in terms of inner representations (or *mental representations*). To simplify somewhat, according to representationalism, for someone to be in a given mental state is a matter of their having the right sort of representation inside their head. For example, to believe that there is tea in the tea caddy is to have an inner representation with the content *there is tea in the tea caddy*. Other mental states are understood in a similar manner. For example, to *desire* that there is tea in the tea caddy—or *hope* or *doubt* or *fear* that there is— each involves possessing a mental representation with the same content. The difference between these mental states is typically understood in terms of the different causal role that the relevant mental representation plays in a person's inner processes. For example, the mental representation associated with believing that there is tea in the tea caddy might cause me to head to the kitchen to make a cup of tea, while that associated with doubting that there is tea in the tea caddy might cause me to call at the corner shop on the way to work instead.

12 MIND AS METAPHOR

Representationalism is commonly applied to beliefs, desires, and other propositional attitudes—such as hoping, wishing, or fearing that such-and-such is the case. It is also applied to other aspects of mental life as well. For example, ideas are often said to be mental representations that arise in our minds in certain circumstances and can be separated or combined in certain sorts of ways. Likewise, concepts are taken to be important forms of mental representations that allow us to categorize objects in our experience and act as building blocks for our thoughts. Mental processes like reasoning, imagining, or planning, are typically thought to involve having a succession of different mental representations. For example, thinking through a maths problem might involve having a series of mental representations about numbers that follow along in a sequence, one after another. Planning a trip to the supermarket might mean running through mental representations of items you need to buy, like milk, bread, or pasta. Imagining going on holiday might involve a series of mental representations about eating ice creams, relaxing on the beach, having a refreshing dip in the pool, and so on.

The representational theory of mind has a long history. It is often traced back to Aristotle and different versions of the approach are attributed to Hobbes, Descartes, Locke, Berkeley, and Hume (Haugeland 1985; Cummins 1989). Different versions of representationalism describe mental representations and their interactions in different ways. Some (like Berkeley or Hume) focus largely on ideas and take them to be akin to inner pictures or images. Others (like Hobbes) think of mental representations as inner symbols, rather like language or mathematical notation. Representationalists also differ markedly in their underlying conception of the metaphysics of mind. Some (like Hobbes) are *materialists* who claim that mental representations are physical entities, typically assumed to be located in the brain. Others (like Descartes) are *dualists*, who argue that thoughts and ideas exist in an entirely different kind of substance to that of the physical world. Finally, some (like Berkeley) are *idealists*, viewing the physical world itself as somehow a construct of our ideas (or ideas in the mind of God). Despite their considerable disagreements, however, all these thinkers share the fundamental vision of the mind as a world of inner representations.

As well as its distinguished history, the representational theory of mind remains the dominant conception of the nature of the mind today. Indeed, it is widely taken to be a key part of the foundations of cognitive science. At the heart of much work in cognitive science is the *computational theory of mind* (or *computationalism*). There are a variety of different computational approaches. In its canonical form—perhaps most often associated with the work of Jerry Fodor (e.g. 1975, 1987)—computationalism accepts the central tenet of representationalism, that the mind can be understood as an inner realm of representations. It then adds the further claim that thinking involves computational processes that operate on these representations. To understand the main idea here, notice that some operations that we carry out upon representations depend only upon their *syntactic* properties (that is, their *form* or shape), rather than their *semantic* properties (that is, their *meaning*). For example, if we're studying logic we know that from $p \rightarrow q$ and p we can infer q, whatever the content of p and q happens to be. Our manipulation of these symbols depends only upon their form, not their meaning. And yet, if we follow the rules correctly, our operation will also make sense when we consider its meaning. For example, if you know that if I am writing, I drink tea and you know that I am writing, then you'll conclude that I'll drink tea. In this way, syntax can 'mirror' semantics. The development of computers in the mid-twentieth century showed that operations like these, which depend only upon the syntax of representations, could be implemented purely mechanically—by using electrical currents as symbols, say, rather than marks on paper.

Like earlier representationalists, computationalists differ in their conception of the nature of mental representations. According to Fodor (1975), mental representations are language-like inner symbols that constitute a *language of thought*. In his view, thinking involves what is now called *classical* computation—that is, the rule-governed manipulation of these inner symbols. In contrast, some computationalists pursue alternatives to classical computationalism that point towards rather different views of the nature of mental representation. For example, *connectionism* tries to make sense of cognition using *artificial neural networks*—collections of interconnected units intended to model the connections between neurons in the brain. Rather than language-like inner symbols, connectionism suggests that mental representations might be distributed across the connections in such networks. Despite these disagreements, however, computationalists are largely united in their commitment to materialism, unlike earlier representationalists. Mental representations and processes are typically assumed to be realized in the brain, much as a computer's data structures and programmes are realized in its hardware. Indeed, much of the appeal of the computational theory of mind—at least in its classical form—is that it promises to offer a thoroughgoing materialist vindication of folk psychology as a theory of our inner machinery. (As we shall see in Section 1.2.2, the relationship between folk psychology and connectionism is more contested.)

Recall two key features of our ordinary talk about thoughts that we noted in Section 1.1: thoughts are described as having *contents* (e.g. being about tea caddies) and as figuring as *causes* (e.g. as causing someone to go to the shop). In its classical version, computationalism seems to offer a way to interpret both of these features of folk psychology as largely *true* claims about our inner workings. For computationalism claims that people *do* have inner representations that express the contents of their beliefs and desires. Roughly speaking, if I believe there is tea in the caddy then I have an inner representation that means *there is tea in the caddy*. This representation is realized by a state of my brain, much in the same way that the sentences of this book are realized by my computer's hard drive. Moreover, computationalism shows how these inner representations can *cause*

14   MIND AS METAPHOR

other inner representations in a way that is entirely mechanical and yet respects their meaning. For example, if my brain is structured in the right way, then an inner representation with the content *if I am writing, I drink tea* together with another inner representation with the content *I am writing* might cause a third with the content *I'll drink tea.*

This is all much too simplistic, of course. In reality, the computational theory of mind is far more sophisticated than this brief sketch might suggest. And yet, despite its sophistication and undeniable appeal, computationalism still faces a number of important challenges. First and foremost, it has yet to explain how mental representations gain their *content.* How on earth can one part of the physical world (e.g. a brain state) come to represent another object or event in the world (e.g. a tea caddy or the corner shop on Union Road)? We shall return to this issue in Chapter 3. For now, let us turn to consider the other possible approach to folk psychology that we mentioned earlier: eliminativism.

1.2.2 Eliminativism

According to the representational theory of mind, folk psychology is a theory about what goes on inside our heads and this theory is, by and large, true. *Eliminativism* agrees that folk psychology is a theory about our inner workings but claims that this theory is false. The folk are trying to describe what goes on inside our heads but, unfortunately, they have got it wrong. And so, the eliminativist says, we ought to conclude that mental states—beliefs, desires, hopes, fears, and the rest—do not exist. To put it bluntly, people do not have minds. The folk might have thought that they did, but it turns out that they were mistaken. This might seem hard to accept at first. Still, eliminativists will remind us, we once believed in the existence of witches, the ether, and crystalline spheres. At some point, we simply had to give up these cherished ideas—and it was a good thing that we did too.

Why do eliminativists claim that folk psychology is false? Some argue that it is not an especially successful theory. For example, Paul Churchland (1981) claims that folk psychology cannot explain many aspects of human behaviour, such as sleep, creativity, or mental illness, and that it has made little progress over the last 2,500 years or so. Eliminativists also argue that folk psychology fits poorly with the theories found in our latest cognitive science, especially neuroscience. To make their case, eliminativists must first say precisely *what* claims they think folk psychology makes about our inner machinery—before showing how these claims somehow fall foul of developments in cognitive science. In this vein, Churchland (1989) claims that folk psychology is committed to the idea that beliefs and desires are inner sentences—that is, internal representations with a language-like structure. As we have seen, Fodor argues that this idea is borne out by the classical computational theory of mind. In contrast, Churchland is more impressed by connectionist networks and argues that these networks involve representations that are radically unlike language. As a result, he concludes that folk psychology is false. (For related arguments, see Patricia Churchland 1986 and Ramsey, Stich, and Garon 1990.)

These arguments are controversial. On the one hand, some authors will question whether folk psychology in fact makes such specific claims about the form taken by mental representations. For instance, perhaps map-like inner representations would serve the needs of folk psychology just as well (Braddon-Mitchell and Jackson 2007)? On the other hand, some authors will question whether connectionism is the correct approach to the mind or dispute Churchland's claim that it shows that mental representation is non-linguistic. For example, perhaps connectionist networks simply provide an alternative way to implement classical computational processes (Fodor and Pylyshyn 1988)? These issues are complex and often depend upon disputes over the correct technical interpretation of particular theoretical approaches in cognitive science, such as connectionism. Despite their complexity, however, each of these debates shares a common starting point—namely, the idea that folk psychology is an attempt to describe our inner machinery. Let us now consider a different way of understanding folk psychology.

1.3 Mind as Metaphor

The core idea of mental fictionalism is that folk psychology is not a theory of our inner machinery, but a useful set of metaphors for making sense of our behaviour. Fictionalism is therefore an *anti-Cartesian* approach to the mind, in the rather broad-brush sense of the term that I introduced earlier. To develop this idea, I shall draw on a particular approach to metaphor and figurative language due to the philosopher of art Kendall Walton. In Section 1.3.1 I will give a brief introduction to Walton's approach. In Section 1.3.2 I will say how I think we can use his analysis of metaphor to make sense of talk about the mind.

1.3.1 Metaphor and Make-believe

Walton's analysis depends upon the idea of a game of make-believe. Imagine some children playing with a doll. They pick the doll up and hold it in their arms, push it down the hallway in a pram, cover it with a blanket, and so on. As they do so, the children imagine themselves picking up a baby and cradling it, taking it to the shops, and lying it down in its cot to go to sleep. In Walton's terminology, dolls and other objects used in make-believe are called *props* and the rules that govern

16   MIND AS METAPHOR

their use in the game are called *principles of generation* (Walton 1990). Within a game of make-believe, the properties of the props, together with the relevant principles of generation, make propositions *fictional*. To say that a proposition is fictional, in Walton's sense, is simply to say that participants in the game are prescribed to imagine it. For example, if the doll's 'eyes' are closed, the children are supposed to imagine that the baby is asleep. It is fictional that the baby is asleep. Notice that, since the content of a game of make-believe depends only on the props and principles of generation, it possesses a certain kind of 'objectivity': if the doll's eyes are closed then it is fictional that the baby is asleep, even if none of the children happen to notice this.

One important feature of games of make-believe is *participation*. Children playing with a doll do not simply sit and look at it. Instead, they carry out various actions with the doll (e.g. putting a bottle to its 'mouth') and these actions generate imaginings within the game (e.g. that the children are feeding the baby). Like the doll, the children themselves become props in the game. The children also participate verbally. For example, David might sing a lullaby as he rocks the doll in his arms, thereby making it fictional that he is singing a lullaby to get the baby to sleep. Importantly, acts of pretence can be used to make genuine assertions. Suppose that Anna looks at the doll and says, 'She's sleeping now!' When she says this, Anna pretends to assert that the baby is asleep. And yet she also indicates that pretending in this way is appropriate. In doing so, Anna makes a genuine assertion: she claims that the props are in a particular state—the state such that it is appropriate to pretend in the way that she does. In other words, Anna claims that the doll's 'eyes' are closed. Once again, notice the objectivity of games: if, in fact, the doll's 'eyes' are open, then Anna's pretence is inappropriate and her assertion is false.[1]

The reason that children play with dolls, it seems, is not because they are interested in the properties of the doll as a lump of plastic. Dolls are interesting only in so far as they allow the children to immerse themselves in a make-believe world in which they can cradle a baby, take it out in its pushchair, feed it, or rock it to sleep. Walton (1993) calls this kind of make-believe *content oriented*. He distinguishes it from a rather different use of make-believe, which he calls *prop oriented*. Walton's central examples of prop-oriented make-believe are cases of metaphor or figurative language:

Where in Italy is the town of Crotone?, I ask. You explain that it is on the arch of the Italian boot. 'See that thundercloud over there – the big angry face near the horizon,' you say; 'it is headed this way'. Plumbers and electricians distinguish

---

[1] Note that here I adopt a pragmatic, rather than semantic, interpretation of Walton's account (for more on this distinction, see Friend 2007).

MAKING UP MINDS   17

between 'male' and 'female' plumbing and electrical connections. We speak of the saddle of a mountain and the shoulder of a highway.

All of these cases are linked to make-believe. We think of Italy and the thundercloud as something like pictures. Italy (or a map of Italy) depicts a boot. The cloud is a prop that makes it fictional that there is an angry face. Male and female plumbing or electrical connections are understood to be, fictionally, male and female sexual organs. The saddle of a mountain is, fictionally, a horse's saddle. But our interest, in these instances, is not in the make-believe itself, and it is not for the sake of make-believe that we regard these things as props. . . .

Make-believe . . . is useful in these cases . . . for articulating, remembering, and communicating facts about the props – about the geography of Italy, or the identity of the storm cloud, or functional properties of plumbing or electrical fixtures, or mountain topography. It is by thinking of Italy or the thundercloud or plumbing connections as potential if not actual props that I understand where Crotone is, which cloud is the one being talked about, or whether one pipe can be connected to another.   (Walton 1993, 40–1)

In the children's game, the doll is a prop. The properties of the doll (e.g. that its plastic 'eyes' are closed, that David is holding it) together with the rules of the game, make certain propositions fictional (e.g. that the baby is asleep, that David is rocking it). According to Walton, the same basic structure underlies many instances of metaphor and figurative language. When we speak of Italy as a boot, the geography of Italy becomes a prop in a game of make-believe. The properties of Italy (e.g. that Crotone sits on the Gulf of Taranto, that Rome lies halfway up its western coast) together with the rules of the game, make propositions fictional (e.g. that Crotone is on the arch of the boot, that Rome is on the shoelaces).

The key difference between the children's game and the Italy-as-a-boot game lies in the interests we have in playing these games. For the most part, the children are not interested in the properties of the doll for its own sake. The doll's properties (e.g. whether its 'eyes' are closed) are interesting only because they shape the content of the fictional world (e.g. whether the baby is asleep). For this reason, the children's game is content oriented. By contrast, when we play the Italy-as-a-boot game, we *are* interested in the properties of our prop—that is, the geography of Italy. The content of our fictional world (e.g. whether Crotone is on the arch of the boot) is interesting only because it helps us to understand our prop (e.g. where to find Crotone on the Italian coastline). The line between content-oriented and prop-oriented make-believe need not be hard and fast, however (Walton 2000, 94). After all, our interests can change from time to time. A child playing with a doll might pause to tell us that its 'eyes' are made from glass marbles. For a moment, her game might become prop oriented. Alternatively, we

18   MIND AS METAPHOR

might find ourselves daydreaming about the Italian boot, imagining it kicking Sicily across the Mediterranean. In this way, the Italy-as-a-boot game might become content oriented, at least for a second or two.

Despite the differences between content and prop-oriented games, Walton analyses the things that we say when we play them in a similar manner. Suppose that John says, 'Crotone is on the arch of the Italian boot'. In doing so, John engages in pretence, much as Anna does when she looks at the doll and says, 'She's sleeping now!' Of course, the pretence involved in the Italy-as-a-boot game is less lively than that found in children's games. It may be that John merely indicates the relevant pretence, without properly engaging in it (Walton 1990, 1993). The key point is that, just like Anna, John indicates that pretending in this way is appropriate. In doing so, John makes a genuine assertion: he claims that the props are in a particular state—the state such that it is appropriate to pretend in the way that he does. In other words, John asserts that Crotone is in such-and-such a position on the Italian coastline. Invoking the familiar game of make-believe in which Italy is imagined to be a boot provides him with a more colourful and memorable way of communicating this fact.

In cases of metaphor, we talk about one topic (our *primary domain*) in terms taken from another area (our *secondary domain*). In Walton's analysis, it is pretence that brings these domains together. The primary domain is that of our props; the secondary domain is the content of the make-believe. Suppose we say, 'Ruth is at a crossroads.' Here we invoke a familiar game in which we pretend that events in someone's life (e.g. important decisions, difficulties they encounter) are stages in a journey (e.g. crossroads, steep paths to be climbed). In Walton's terminology, the events in their life (i.e., our primary domain) are props, and the stages in the journey (i.e., our secondary domain) are the content of the make-believe. Our utterance is an act of pretence within this game. In this context, when we say 'Ruth is at a crossroads' we do not claim that she is literally standing at a road junction. We only pretend to assert this, just as Anna only pretended that the baby was asleep, and John only pretended that Italy was a boot. And yet, much like Anna and John, we also indicate that our pretence is appropriate and thereby make a genuine assertion about our props: we claim that Ruth must make an important decision. It is this fact about Ruth's life that makes our pretence appropriate within the game. If Ruth were not faced with an important decision, our pretence would be inappropriate, and our assertion would be false.

1.3.2 Myths and Metaphors

We can now develop a fictionalist analysis of folk psychology. Let us begin with a famous myth about the origin of our talk about mental states, due to Wilfrid Sellars (Sellars 1956; cf. Yablo 2005). Sellars asks us to imagine a society that at first uses only what he calls a 'Rylean' language that is restricted to terms referring to overt behaviour. At some point, along comes a visionary theorist, called Jones, who develops a theory of internal, psychological episodes, which he dubs *thoughts*. Jones bases his theory of thoughts on the model of overt verbal behaviour. However, like all models, Jones' model is accompanied by a commentary specifying which of the features of its source domain (overt verbal behaviour) are to be carried over to its target domain (internal psychological states). For example, Jones is clear that his theory doesn't claim that there is any mysterious inner tongue that utters thoughts, or that thoughts can be heard. Using his theory, Jones is then able to predict and explain the complexity of people's behaviour in a way that the original Rylean language could not.[2]

Sellars' myth is often cited in support of the idea that folk psychology is a kind of proto-scientific theory of our inner machinery. I want to propose an alternative, fictionalist myth about the origin of folk psychology. In the fictionalist version, Jones is not a visionary theorist. Instead, he is rather like the first person to describe Italy as a boot. What Jones introduces to our Rylean ancestors is not a new theory of an inner realm, but a useful metaphor. Jones suggests that we should talk about people *as if* they undergo inner episodes that are analogous to overt verbal behaviour. Like overt verbal behaviour, these inner episodes can represent the world: they have contents and can be true or false. They can also respond to experience, interact in certain ways, and cause certain behaviours. According to this alternative myth, Jones still takes overt verbal behaviour as his model for thoughts. But the commentary he offers on this model is very different from that envisaged by Sellars. In the fictionalist telling of the tale, Jones is no more committed to the existence of these inner episodes than he is to the existence of an inner tongue. Instead, the entire model is proposed merely as a useful metaphor for describing people and their behaviour. Thoughts are not theoretical entities, but useful fictions.

For the fictionalist, the moral of the story is clear. We talk about thoughts *as if* they were inner states that represent the world and cause people to behave in certain ways. According to the fictionalist, this conception of thoughts is fundamentally metaphorical. To use the terminology I introduced in Section 1.3.1, our *primary domain* is people and their behaviour. This is what we want to talk about and understand. Our *secondary domain* is the realm of external, public representations, especially spoken and written language. This is what we use to make sense of our primary domain. Our conception of thought is a metaphorical mapping of the outer world of public, external representations onto a private, internal world of

---

[2] Sellars (1956) refers to Jones' thoughts as 'inner speech'. I followed this use of the phrase in Toon (2016) but have avoided it here. As Sellars acknowledges, the phrase 'inner speech' can be misleading. It is often used to indicate a particular sort of sensory experience—namely that of 'hearing' our own voice speaking to us 'in our heads'. The relationship between our thoughts and this experience of 'hearing' inner speech is complicated. We will consider this issue in Chapter 2 (Section 2.3.2).

20 MIND AS METAPHOR

the mind. We talk about the mind *as if* it were a private, inner world of representations that express our thoughts—a hidden realm of representations that have content or meaning, can be true or false, accurate or inaccurate, and so on. This inner world does not exist. It is merely a useful fiction. And yet our metaphors are invaluable to us. They are invaluable because they *work*: they help us to pick out genuine patterns in people's behaviour (including our own). As we will see, however, the reason that these metaphors work—when they do work and to the extent that they do—is simply because the patterns of behaviour that we exhibit when we are *not* using external representations are often similar to those we exhibit when we *are* using them (see Chapter 4).[3]

Sellars' original myth is concerned with thoughts as *episodes*—i.e., thoughts that, like overt speech, are said to take place at a particular time. As we have seen, I am interested in thoughts in a broader sense, including different sorts of propositional attitudes. Only some of these are described as events or episodes, which take place at a particular time. Good examples are *decisions* and (what I will call) *judgements*. If I decide to make a cup of tea, then it seems that I do so at a particular time. Likewise, if, in a moment of quiet reflection, I come to the shocking realization that I drink too much tea, my judgement that I do so would seem to take place at a particular (and rather sad) moment in my life. Other propositional attitudes are longer lived. Consider intentions and beliefs. It seems odd to say that these take place at a particular time. After all, I might intend to make a cup of tea all morning. Similarly, I might have the belief that tea tastes good for days, weeks, or months—or even my whole life. Sometimes, this distinction is marked by talking about *occurrent* thoughts and *standing* states. For example, my sudden realization that I drink too much tea might be described as an occurrent belief, while my belief that tea tastes good might be described as a standing belief. This distinction is often taken to be related to conscious awareness: we are said to be immediately aware of our occurrent beliefs, in a way in which we are not (or, at least, are not always) aware of our standing states. Later, I will argue that this is a mistake (see Chapter 2). For the moment, the important point is simply that there are different kinds of propositional attitudes, which we talk about in different ways. (To avoid confusion, I will stick with the term 'judgement' and use 'belief' only for standing states.)

Different sorts of thoughts are best served by different sorts of metaphors. The world of external, public representations is enormously varied: it contains many different sorts of representations used in many different sorts of ways. As a result, it offers a vast stock of metaphors for talking about the mind. In our re-telling of

---

[3] These points are intended to elaborate on the initial formulation of the fictionalist approach I gave in Toon (2016). I now realize that I needed to be clearer about the nature of the primary and secondary domains involved in folk psychology, and the relationship between them. For helpful and perceptive critical discussion on this point, see Bourne and Caddick Bourne (2020) as well as Ross (2022).

Sellars' myth, the source for Jones' metaphor was overt verbal behaviour. Our verbal behaviour takes many different forms, of course. We make assertions, ask questions, discuss things, make promises, offer apologies, put forward arguments, and much else besides. Each of these sorts of verbal behaviour lend themselves to different metaphorical purposes. The same is true when we turn to consider other forms of external representation, especially written language. Despite its variety, verbal behaviour is typically fleeting. A sentence is uttered in one moment and lost in the wind the next. By contrast, written language is longer lasting and is put to different uses as a result. This also means that it lends itself to different metaphorical uses in our talk about the mind. Let us turn to some examples.

Consider decisions. Some decisions are taken out in the open. Suppose that a committee needs to decide what to do about a city's traffic problems. The committee meets and deliberates for a while, talking through possible courses of action—say, whether to build a new road or improve the train service. At some point, the committee states their decision, which is recorded in the minutes. Once it is recorded, the decision shapes the committee's future actions. For instance, if they've decided to improve the train service, they might need to invest in new rolling stock or raise fares. When an individual makes a decision, we often talk as if they had undergone a similar process, albeit one that is conducted by an unseen committee, operating behind the scenes. We talk as if they had silently debated the matter with themselves, weighing up different options available to them, before finally reaching a conclusion and entering it into the record to guide their future actions. Or consider judgements. Here our model is something like a courtroom. The judge weighs up the evidence, deliberates in her chambers, then pronounces her judgement—say, that the defendant is guilty. The judgement is duly noted by the court stenographer and entered into the public (and defendant's) record. Once again, we talk as if, when someone makes a judgement, the same sort of thing happens in a closed courtroom that sits inside their head: they assess the evidence, pronounce their judgment, and this is duly entered into their memory for future use.

Both these metaphors look mainly to spoken language: a process of deliberation takes place, and the verdict is declared. By contrast, the metaphors we use for standing states tend to be drawn from uses of written language. Consider belief and memory. If we want to record some important information, we can jot it down in a notebook. Later, when we want to use this information, we can look it up in the notebook, and act accordingly. According to the fictionalist, our ordinary concept of memory is based upon practices like these, in which we use items of material culture to record information. We often talk about memory as if it were a kind of private, inner notebook that we keep tucked away inside our heads. As someone goes about the world, we can imagine them jotting down important information in their inner notebook. Later, when they need this information, we can imagine them looking it up in their inner notebook and

22 MIND AS METAPHOR

acting accordingly. In this way, the fictionalist suggests, we draw upon our practices for using certain forms of external representation—in this case, notebooks—as a metaphor for making sense of people and their behaviour. The metaphor works because people *do* behave somewhat as if they had a little notebook inside their heads. In other words, there are similarities between the way people behave without notebooks and the way that they behave with them. (Of course, these metaphors can also mislead us. We will return to this point in Chapter 4. For a history of the different metaphors that have been applied to memory and their pitfalls, see Draaisma 2000.)

Other forms of material culture provide important inspiration besides notebooks. Think of an architect's drawings (cf. Houghton 1997). We call such drawings 'plans', of course. And it is precisely external, material representations like these— along with other documents that fulfil similar functions, like route maps for upcoming journeys or itineraries for day trips—that provide our means for talking about what people do when trying to achieve certain goals. Suppose that Ruth wants to put up some shelves in the living room. How might she go about it? She might write down a list of what she needs to do (go to the shed, fetch the toolbox, mark out screw holes, etc.) and draw a sketch of how she wants to position the shelves. Alternatively, she might manage without pencil and paper. Even if she does, however, we can make sense of her behaviour (her trip to the shed, her drawing pencil marks on the wall in such-and-such a way, etc.) by pretending that she does have such a list and sketch guiding her actions. Similarly, we can think of someone's desires as if they were a kind of private, inner shopping list (or 'wish list') that they will try to tick off when they get the opportunity. (Later on, we will explore some of these metaphors in more detail, along with further examples.)

Once we recognize the role of metaphors in our talk about the mind, we can start to develop a new approach to our attributions of mental states. To do so, we can make use of Walton's analysis of metaphor. Recall that, according to Walton's analysis, a metaphor's primary and secondary domains are brought together through pretence. When we say, 'Ruth is standing at a crossroads', we invoke a well-known game of pretence that treats life as a journey. In doing so, we give ourselves a convenient, and perhaps more vivid, way of describing the events in someone's life. We talk about their life by making moves in this game. When we say, 'Ruth is standing at a crossroads', we are not making a straightforward assertion; our utterance is an act of pretence. And yet, by pretending in this way, we indicate how to pretend correctly within the life-as-a-journey game. In doing so, we also make a genuine assertion about Ruth and her life. What makes our pretence appropriate (or inappropriate) and our assertion true (or false) are the facts about events in Ruth's life—in particular, whether she finds herself faced with an important decision or not.

The metaphors that we use to talk about the mind can be understood in a similar way. For instance, by invoking a game of pretence that treats memory as an inner notebook, we give ourselves a convenient—indeed, as we shall see, an indispensable—way of describing someone's behaviour. We talk about their behaviour by making moves in this game. For example, suppose we say that Mark believes that the No. 73 bus goes to Oxford Street. When we say this, we are not claiming that he has a representation with this content lodged somewhere inside his head; we are indicating how to pretend correctly within the memory-as-a-notebook game. In doing so, we also make a genuine assertion about Mark and his behaviour. We are saying that Mark is in a certain state—the state such that it is appropriate to pretend in the way that we do. What makes our pretence appropriate (or inappropriate) and our assertion true (or false) is a complex set of facts about Mark's behaviour. For example, when he wants to go to Oxford Street, he jumps on the bus marked '73'; if someone at the bus stop asks him if the No. 73 goes to central London he says 'yes'; and so on. If, on the other hand, he had confidently jumped on the 254 or answered 'no' to their question, then our pretence would have been inappropriate and our assertion false. Put simply, Mark behaves *as if* he had 'the No. 73 goes to Oxford Street' written in his inner notebook—even though there is no such notebook.

For the fictionalist, then, folk psychology is not a theory of an inner world. Instead, it is a game of prop-oriented make-believe. We use this game to make sense of people and their behaviour, just as we use the life-as-a-journey game to make sense of people's lives, or the Italy-as-a-boot game to make sense of the geography of Italy. The props in our game are people and their behaviour. Its principles of generation are the complex, nuanced, and largely tacit rules that govern our folk psychological practices. Together, these props and principles of generation generate the content of our make-believe—an imagined, hidden realm of inner deliberations, verdicts, notebooks, arguments, calculations, plans, and shopping lists. We pretend that people possess these inner worlds, just as we pretend that the events in someone's life are stages in a journey, or Italy is a boot. Our ordinary talk about the mind involves acts of pretence within our folk psychological game. By pretending to say that Crotone is on the arch of a giant boot, we manage to say something true about its location on the coastline of Italy. In a similar manner, by pretending to say that people harbour inner representations, we manage to say something true about their behaviour.

Earlier, in Section 1.3.1, we saw that the distinction between content and prop-oriented make-believe is not a sharp one. It is a matter of where our interests lie, and these can vary at different times and in different contexts. In Walton's analysis, metaphors usually involve prop-oriented games. When we use a metaphor, our interest lies first and foremost in the objects we are talking about, not the pretence we are invoking. We pretend that Italy is a boot in order to understand its geography, not for pleasure of contemplating a giant boot floating in the Mediterranean. There is nothing to stop us daydreaming in this way if we so wish, however. If we do, the Italy-as-a-boot game will become content oriented, at

24 MIND AS METAPHOR

least as long as our daydream lasts. Most of the time, it would seem, our folk psychological game is also prop oriented: our interest lies in describing people and their behaviour, not daydreaming about a world of sentences or pictures inside their head. There are times when our folk psychological game might become content oriented, however. This often happens when we are doing philosophy: instead of putting the make-believe to practical use to understand its props (i.e. people and their behaviour), we begin to focus instead on its content (i.e. the nature of the inner world that it invokes). Like daydreaming about a giant boot, this is harmless enough in itself. The trouble comes only if we begin to take our daydreaming (or philosophizing) too seriously.

1.4 Implications

Section 1.3 introduced the main idea behind a fictionalist approach to the mind. There is much that needs to be done to refine and develop this approach. I shall take up this task in later chapters. First, let us pause to consider how fictionalism relates to the other approaches that we encountered earlier, namely representationalism and eliminativism. As we saw, folk psychology is usually seen as a theory of our inner machinery, a view I called Cartesianism. Representationalists claim that this folk theory is true. Eliminativists claim that it is false and conclude that mental states do not exist. How does fictionalism differ from these views? It is helpful to distinguish two issues here. First, how does fictionalism interpret folk psychology? Second, what does it say about the existence of mental states?

1.4.1 Making Sense of the Folk

The fictionalist analysis of folk psychology contains two key claims. First, that folk psychology involves the notion of the mind as an inner world. Second, that this notion is merely figurative. Cartesians—both representationalists and eliminativists alike—accept the first claim. Indeed, it is often simply taken for granted that the folk regard mental states as inner states. There are anti-Cartesian approaches to the mind that reject this claim, however, such as behaviourism and instrumentalism. We shall consider how fictionalism relates to these views in Chapter 2. For now, let us accept that idea that the folk conceive of the mind as an inner world and ask what attitude they take towards this idea. According to representationalism and eliminativism, the folk take it literally: it is a theory of what is going on inside our heads. In contrast, the fictionalist claims that such talk is merely figurative. In particular, I have suggested that it is metaphorical and that we can understand these metaphors in terms of pretence.

At first glance, the fictionalist's analysis might seem incredible. After all, we do not ordinarily feel as if we are speaking metaphorically, much less pretending, when we say what people think or want or feel. If we say that Mark believes the No. 73 bus goes to Oxford Street, we do not feel as if we are engaging in make-believe; we feel as if we are making a straightforward assertion about Mark. How then can fictionalism claim to give a plausible interpretation of our ordinary talk about the mind? This worry is often called the *phenomenological objection* to fictionalism (Wallace 2022; see also Eklund 2019).

The first point to stress is that, according to the fictionalist, we *are* making a genuine assertion when we attribute a mental state to someone. When we say 'Crotone is on the arch of the Italian boot' we are making a genuine assertion about its location. We simply do so by means of the pretence that Italy is a boot. Similarly, when we say 'Mark believes that the No. 73 bus goes to Oxford Street' we are making a genuine assertion about Mark. We simply do so by means of the pretence that his thought is captured in an inner sentence. The fictionalist would also expect that, when we talk in this way, our overriding impression is that we are making a genuine assertion, rather than engaging in pretence. After all, it is characteristic of prop-oriented make-believe that our attention is fixed upon objects in the real world, rather than our make-believe. If you and I had a heated dispute about whether Crotone is on the laces or the arch of the Italian boot, our main concern would likely be who was right or wrong about the location of Crotone. Our focus would be on *what* we were saying, not *how* we were saying it. The same is true of ordinary talk about the mind.

So the fictionalist need not deny that talk about the mind involves genuine assertions, nor even that this is how it typically feels to us. And yet surely, the critic will object, it seldom feels as if we are speaking metaphorically or engaging in pretence *at all* when we attribute mental states—even metaphor or pretence that is in the service of serious assertion. Contrast this with our talk about Italian boot. However caught up we might be in our argument about Crotone, we are surely still at least dimly aware that we are speaking figuratively when we insist that it is on the shoelaces or the arch of a boot. Can the same be said of our ordinary talk about the mind? Are we even dimly aware that such talk is metaphorical? If we're not, how can the fictionalist maintain that such talk is metaphorical after all?

The issues here are more subtle than they might first appear. Often, we do not notice when we are speaking metaphorically, especially when metaphors are familiar to us (Yablo 2000; see also Eklund 2019). Consider a famous example given by Lakoff and Johnson (1980). We often speak of arguments as if they were battles: we talk of *defending* a theory, *winning* or *losing* an argument, *shooting down* a claim, trying a new *strategy*, judging whether we are *gaining ground*, and so on. When we talk in this way, we usually pay little or no attention to the fact that we are speaking metaphorically. Indeed, the metaphorical character of such talk might never even have occurred to us before someone points it out. Other

26 MIND AS METAPHOR

metaphors are even more subtle and difficult to spot. Consider the way we talk about time in terms of space: we say we are *looking forward* to a holiday, that the future *lies ahead of us*, that we want to put the unpleasant argument *behind us*, that we must *face up* to what is *coming*, and so on. These ways of talking are so deeply ingrained in our language that it takes a while to recognize their metaphorical character, even once this is pointed out to us.

Indeed, it is generally the case that engaging in some domain of discourse is a different matter to offering an analysis of it. Ryle compares the ordinary speaker to a villager who knows his local area. The villager might know perfectly well how to get from place to place and yet still struggle if asked to produce a map of the area. By contrast, the philosopher's task, like that of the cartographer, is precisely to produce such a map of the territory. It should not be too surprising if the ordinary speaker's workaday knowledge and the philosopher's more abstract knowledge do not immediately tally with one another: although they might both understand a particular area of discourse, their understanding is of a different sort. How can we know whether our philosophical map is accurate or not? One way to check its accuracy, Ryle tells us, is to see whether it leads us to absurdity. This is Ryle's central objection to Cartesianism, of course. If we mistakenly place mental conduct terms in the categories of non-physical 'thing' or 'stuff', then we are led to ask seemingly unanswerable questions, like 'how can the mind interact with the body?', 'how can we refer to things inside people's heads?' or 'how can we ever access the minds of others?'

For Ryle, the notion of the mind as an inner world is imposed upon ordinary language by philosophers. For the fictionalist, such talk is part of ordinary language, but it is merely figurative. Despite their differences, however, the fictionalist can still adopt Ryle's strategy for revealing the nature of folk talk about the mind. Like Ryle, the fictionalist will argue that, if we insist on taking the idea of an inner realm seriously, then we are led to absurdity. The fictionalist's diagnosis of the situation is rather different, however. The absurdity that Ryle so brilliantly teases out of Cartesianism is not a sign that Descartes has got his philosophical map wrong—or, at least, not entirely wrong. The folk *do* talk about mental states as if they were inner states. The absurdity results from taking such talk too literally. Ryle famously brands Cartesianism a category mistake. But category mistakes are often an indication that we are dealing, not with a misguided theory, but with a metaphor. Indeed, Paul Ricœur remarks that 'it is tempting to say that a metaphor is a kind of planned category mistake' (1978, 233; see also Magidor 2022).

Consider the way that Ryle pokes fun at epistemologists' descriptions of intellectual activity in terms of supposed inner cognitive acts or processes, using terms like 'judgement', 'reasoning', 'concept', 'idea', 'inferring', 'drawing conclusions from premises', 'considering propositions', 'intuition' (1949, 275). Ryle writes that

Such expressions are employed, not indeed by the laity but by theorists, as if with their aid, and not easily without it, correct descriptions can be given of what has at a particular moment been occupying a particular person; as if, for example, John Doe could and should sometimes be described as having woken up and started to do some judging, conceiving, subsuming, or abstracting; as spending more than three seconds in entertaining a proposition, or in moving from premises to a conclusion; or as sitting on a fence, alternately whistling and deducing; or as having had an intuition of something a moment before he coughed.

Probably most people feel vaguely that there is a tinge of unreality attaching to such recommended biographical anecdotes. John Doe's own stories about himself are not expressed in such terms, or in terms easily translatable into them. How many cognitive acts did he perform before breakfast, and what did it feel like to do them? Were they tiring? Did he enjoy his passage from his premises to his conclusion, and did he make it cautiously or recklessly? Did the breakfast bell make him stop short halfway between his premises and conclusion? Just when did he last make a judgement, or form an abstract idea, what happened to it when he had made or formed it and who taught him how to do it? ... He does not know how to begin to answer such questions.   (1949, 275–6)

According to Ryle, terms like 'inference', 'concept', and 'judgement' properly apply not to the *process* of thinking, but to its *products*—such as published theories, lessons given on a blackboard, or a 'detective's typewritten report' (1949, 276). It is here that we find inferences, concepts, judgements, and the rest. Unfortunately, epistemologists have misapplied these terms to describe supposed inner acts or processes, which the scientist, teacher, or detective is taken to undergo while producing their theory, lesson, or report. This is the sort of category mistake that is typical of Cartesianism. As Ryle puts it, 'biographical anecdotes told in these idioms are myths, which means that these idioms, or some of them, have their proper applications, but are being misapplied' (1949, 276). (We will return to Ryle's discussion of the intellect in Chapter 5.)

The fictionalist will dispute Ryle's claim that this conception of intellectual activity is entirely foreign to ordinary language. While some of his examples are largely technical terms (e.g. 'subsuming'), others are surely used in this manner outside of epistemology (e.g. 'reasoning', 'intuition', 'concept'). The fictionalist can agree, however, that we feel a 'tinge of unreality' when describing our inner life in these terms. We would feel a similar tinge of unreality if asked about the causes of the cloud's anger or what breed of sleeping dog we were letting lie. In each case, the reason that such questions baffle us is not because someone has imposed a category mistake upon us, but because we are speaking metaphorically. We often describe intellectual activity *as if* it involved assenting to inner sentences, moving through the steps of an inner argument, holding a silent debate between

28 MIND AS METAPHOR

competing views, and so on. Most of the time we pay little attention to the question of what it is that we are doing when we talk in this way. It is only when someone excessively literal-minded, like a philosopher, demonstrates the absurdity of taking such talk too seriously that we realize that this is not how we normally use these terms.

Perhaps some people will be willing to accept that some, or even most, of our talk about thinking is metaphorical. In many ways, this might not even be terribly surprising. After all, a lot of our talk about the mind is unashamedly figurative. As we noted in the Introduction, everyone will be happy to say that we are speaking metaphorically when we say that we're making a mental note or putting an unpleasant thought to the back of our mind. But what about apparently straight-forward attributions of mental states, like 'Mark believes that the No. 73 goes to Oxford Street' or 'Ruth wants a cup of coffee'? Is there even a hint of metaphor here? Such attributions are certainly less noticeably figurative than talk about reasoning as a silent argument or ideas as inner pictures. And yet the fictionalist will insist that, once we reflect more closely on our talk about mental states, we find that it bears striking similarities to metaphorical language. To see this, consider some well-known, and apparently puzzling, features of our ordinary conception of mental states.

It is often noted that beliefs, desires, and other mental states do not behave like ordinary physical objects. For example, it seems odd to say that beliefs have properties like location, mass, or temperature. How wide is Mark's belief about the No. 73 bus? Is it at the front or back of his head? Is it to the right or left of his desire to have a cheese sandwich? Does it weigh more or less than one gram? What shape is it? Does it cool down in winter? Or warm up when Mark runs after the bus? These questions strike us as ridiculous. And yet, if mental states were physical states, some of these questions at least ought to make sense, even if we cannot answer them. Of course, one response to this worry is to turn to dualism. If mental states are made from a strange, non-physical substance, then we should hardly expect them to have properties like length or weight (Haugeland 1985, 36–7; Braddon-Mitchell and Jackson 2007, 5). Even if we think of mental states as non-physical, however, much of our talk about the mind remains puzzling. For instance, mental states have problematic identity conditions. As Dennett has pointed out, 'common intuition does not give us a stable answer to such puzzles as whether the belief that 3 is greater than 2 is none other than the belief that 2 is less than 3' (1987, 55). It seems foolish to try to count mental states. How many beliefs do you have? Do you have more beliefs than desires? Is there a limit to how many beliefs you can have? These questions seem unanswerable.

Each of these puzzling features of our talk about mental states begins to make sense if this talk is metaphorical. A characteristic feature of metaphors is that they invite 'silly questions' (Yablo 1998)—questions that push the metaphor too far by asking about things that it doesn't specify. For example, suppose a friend complains to you that his latest project is a monkey on his back. You immediately express your surprise: 'Really! What species?' Of course, you might have intended your question metaphorically too. Perhaps you were asking in a tongue-in-cheek way how serious the problem is. (I assume that it's worse to have a hefty baboon on your back than a tiny marmoset.) If you weren't speaking metaphorically, however, then you have asked a silly question. Questions about the physical properties of mental states are like that. To ask where beliefs *are* (or what shape they are or how much they weigh or whether they can heat up or cool down) is to ask a silly question—one that misunderstands the metaphorical nature of our talk about mental states. It is rather like being told that someone has a chip on his shoulder and asking if it is on his right or left shoulder. Mental states lack physical properties not because they are made of strange, non-physical stuff, but because they are creations of our metaphors. (In Chapter 4, we will see how this issue arises in debates over the extended mind thesis.)

Other aspects of our talk about the mind also begin to make sense when seen in this light. Consider the puzzles regarding the identity conditions of mental states. As we noted, these puzzles remain even if we regard mental states as non-physical. But if our talk about mental states is metaphorical, then they are easily explained. As Yablo (2000) points out, objects invoked in make-believe also have problematic identity conditions. Imagine being told that you need to face up to the elephant in the room and asking, 'Is the elephant in the room today the same as the one that was here last week?' Again, you might be speaking metaphorically, wondering whether this tricky issue is the same as the one you avoided last week. If not, you have asked a silly question. A similar explanation can be given, I think, for why we are baffled by questions about the identity conditions of mental states, such as Dennett's puzzle about whether your belief that 3 is greater than 2 is the same as your belief than 2 is less than 3. The creations of metaphor are also difficult to count. Imagine being asked exactly how many butterflies you had in your stomach before that nerve-racking job interview. Counting the number of beliefs and desires that you have is perplexing for much the same reason.

It is often difficult to say whether we have a particular belief or not. Consider a well-known example from Dennett (1969, 2013). Imagine a child who, when she is young, learns to say 'Daddy is a doctor.' Eventually, when she is an adult, she comes to a proper understanding of what this means. At what point does the child count as having the belief that Daddy is a doctor? Five years old, perhaps? Or four? Or six? The trouble with asking this question is that it seems wrong to think of it as an all-or-nothing matter. Instead, it seems more plausible to say that at some point as she is growing up—perhaps even for a few years—the child 'sort of' or 'half' believes that Daddy is a doctor. Perhaps when she is four years old, she can tell you that doctors help make people better, but she couldn't tell you the difference between a doctor and a nurse (or a counsellor or physiotherapist). At this age, it seems uncharitable to say she has no belief whatsoever about what her father does

30   MIND AS METAPHOR

for a living. And yet we would probably hesitate before attributing the belief to her in the same way as we would a grown-up. How can we make sense of this? If beliefs are inner representations, either physical or non-physical, then surely you either have them or you don't (Schwitzgebel 2001)?

Again, this aspect of our talk about mental states is exactly what we would expect if they were metaphorical. The aptness of any metaphor is a matter of degree. Consider the metaphor we use when we describe clouds as angry (Walton 1993). There are cases in which everyone will agree that the metaphor is apt (say, standing in the midst of a Force Ten gale) or that it isn't (gazing up at thin wisps of white cloud on a sunny day). In other cases, competent speakers may differ. The same is true of the metaphors that we use to describe the mind. In some cases, all will agree if asked whether these metaphors are apt or not. In other cases, there may be considerable disagreement. Indeed, even an individual speaker may feel uncertain if asked whether the metaphor is appropriate. The case of the girl whose Daddy is a doctor is like that. Consider her situation when she's four. According to the fictionalist, our attributions of belief are guided by the central metaphor of sentences recorded in an inner notebook that guides our actions. In some respects, this metaphor fits her fairly well (e.g. if you ask her what Daddy does, she will reply 'he's a doctor'). In other respects, it fits less well (e.g. if you point to a nurse in her storybook and ask whether that's what her father does, she might falter). Beyond this, there is no further, hidden fact of the matter about whether she *really* believes that Daddy is doctor. That's just not how metaphors work.

All in all, then, the fictionalist's claim that ordinary talk about mental states is metaphorical might not be as outlandish as it first appears. Of course, there is more that must be said in defence of this interpretation of folk psychology. Some of the features of talk about mental states that we have considered are given alternative explanations by other anti-Cartesian approaches to the mind, such as behaviourism and instrumentalism. We shall consider these views in Chapter 2. First, however, let us ask what fictionalism means for the existence of mental states. Put simply, does fictionalism deny that the mind exists?

1.4.2 Does the Mind Exist?

Both representationalism and eliminativism see folk psychology as a theory of our inner machinery. Understanding folk psychology in this way leaves the existence of mental states as a hostage to fortune. Is our folk theory right about what is going on inside our heads or not? Will it be vindicated by future developments in cognitive science? Or will it turn out that we do not have mental states after all? As we have seen, fictionalism rejects this interpretation of folk psychology. According to the fictionalist, folk psychology is not a theory of the inner, but a set of metaphors for making sense of our behaviour. What does this mean for the existence of mental states? Does the fictionalist think we *really* have beliefs, desires, hopes, and fears?

Consider the metaphor of angry clouds again. For the fictionalist, the question 'do mental states exist?' is like the question 'do angry clouds exist?' In both cases, the answer is: in one sense 'yes' and in another sense 'no'. Are there clouds that are literally angry (or happy or miserable)? No. Are there clouds that are quite properly called angry (or happy or miserable) when we're indulging in a particular sort of pretence? Yes. The same lesson applies to talk about mental states. Are there literally representations inside people's heads that express their thoughts? No. Are there 'real patterns' (Dennett 1991b) in people's behaviour that are picked out when we invoke this pretence? Yes. The upshot is that, for the fictionalist, the fate of folk psychology does not depend upon future developments in cognitive science. Whatever scientists eventually discover about our inner machinery, important patterns in people's behaviour will remain and folk psychology will still allow us to make sense of them.

It would be misleading, then, to say simply that fictionalism denies the existence of mental states. *If* mental states are presumed to be inner representations, then fictionalism does indeed deny that such things exist. Our talk about these inner representations is only metaphorical. But we use these metaphors to pick out features of people's behaviour that are perfectly real. In this sense, fictionalism does not deny the existence of mental states. Instead, it offers us an account of what mental states are and how they are picked out by folk talk about the mind. This point can be easy to lose sight of, since the idea that folk psychology is a theory of our inner states is so deeply entrenched. From this perspective, it can seem obvious that to deny the existence of these inner states is to deny the existence of mental states—that to deny that people possess inner representations is to deny that they have beliefs, desires, hopes, or fears. Indeed, it is tantamount to denying the very existence of the mind. This is a mistake, however. It overlooks the possibility that there is another way of understanding folk psychology and the nature of mental states.[4]

Why deny the existence of mental representations? Earlier we saw that eliminativists have offered a number of arguments intended to show that these inner representations do not exist—at least, not in the form described by folk

---

[4] I made this mistake myself in Toon (2016): I wrote as if denying the existence of inner representations meant denying the existence of mental states. In my defence, I should say that this way of talking is fairly commonplace. Still, it is misleading: there are many ways to affirm the reality of mental states, without taking them to be inner representations. Indeed, mental fictionalism, as I understand it, is one such approach: it affirms the reality of the mental if understood as patterns in behaviour, while treating the inner states used to pick out those patterns as fictions. There are other realist and anti-representationalist approaches too. For instance, a dispositionalist might identify mental states with our tendency to behave in certain ways, rather than with inner representations. Dennett's intentional stance is often seen in a similar light. In Chapter 2, I will say why I think we ought to prefer fictionalism to these alternative approaches. For helpful critical discussion of mental fictionalism on this point (including my own approach), see Drayson (2022) and Hutto (2022a).

32 MIND AS METAPHOR

psychology. In Chapter 3, I shall offer my own reasons for doubting the existence of mental representations. For now, however, it is important to notice that, strictly speaking, the fictionalist might remain agnostic over the question of whether we have these inner representations or not. The distinctive feature of fictionalism is that it allows us to grant that, even if these inner states do not exist, we can nevertheless continue talking as if they do. This represents a clear advantage of fictionalism over eliminativism. Abandoning talk about beliefs, desires, and other mental states would require an enormously dramatic and far-reaching transformation in our language. In contrast, fictionalism aims to show that the legitimacy of ordinary talk about the mind does not depend upon beliefs and desires finding a place in future cognitive science.

Why talk about these inner states if they don't exist? According to the fictionalist, such talk shares the advantages of metaphorical and figurative language more generally. Metaphors expand our powers of expression, allowing us to say things that we could not say otherwise (Walton 1993, 2000; Yablo 1998). They introduce a 'framing effect' (Moran 1989; see also Demeter 2013): we are asked to 'see' our primary subject (e.g. Italy, clouds) in terms of a secondary subject (e.g. a boot, emotions) (Beardsley 1962; see also Hills 2017). This yields a host of different cognitive benefits. Metaphorical claims are often especially vivid and memorable, for example, and prompt a range of further inferences about their primary subject. Each of these points holds for the metaphor of the mind as an inner world. In Chapter 2, we will consider some of the benefits that follow from this metaphor. As we will see, many of the problems faced by other anti-Cartesian approaches, such as behaviourism and instrumentalism, stem from a failure to appreciate the importance of this metaphor in our conception of the mind.

1.5 Conclusion

We often think of the mind as an inner world. We talk about mental states as if they were inner states that represent the world and cause us to behave in the way that we do. Philosophers tend to take this idea seriously. In contrast, I have suggested that it is metaphorical. The notion of the mind as an inner world does indeed lie at the heart of folk psychology. But it is not a hypothesis. It is a useful fiction that we use to make sense of people. One of the guiding motivations for fictionalism is the thought that the folk are not in the business of speculating about our inner machinery. In this respect, fictionalism echoes a number of other well-known approaches to the mind, now often taken to be discredited, such as behaviourism. In the next chapter, we will turn to consider these views and their relationship to mental fictionalism.

2

Contrasts, Causes, and Consciousness

Chapter 1 introduced mental fictionalism and showed how it rejects the widespread idea that folk psychology is a theory of our inner machinery. Fictionalism is not alone in rejecting this idea, of course. The same is true of a number of other approaches to the mind, including behaviourism, instrumentalism and Daniel Dennett's well-known work on the intentional stance. In this chapter, I will consider how fictionalism differs from these longstanding anti-Cartesian approaches to the mind (Section 2.1). In doing so, I aim further to clarify and motivate mental fictionalism. A central theme throughout the discussion will be that existing anti-Cartesian views face difficulties because they reject outright the notion of the mind as an inner world. In contrast, fictionalism fares better since it recognizes that this notion plays an important, albeit metaphorical, role in ordinary talk about the mind. Afterwards, I consider two objections that can be levelled against all anti-Cartesian approaches: that they deny the obvious fact that mental states are causes of behaviour (Section 2.2) and that they fail to account for the immediate, conscious awareness that we are often taken to have of our own mental states (Section 2.3). I will try to show how fictionalists can respond to these objections.

2.1 Anti-Cartesianism

Let us consider how mental fictionalism relates to existing anti-Cartesian approaches to the mind, including behaviourism (Section 2.1.1), Ryle's view (Section 2.1.2), instrumentalism (Section 2.1.3), Dennett's intentional stance (Section 2.1.4), and so-called prefix-fictionalism (Section 2.1.5).

2.1.1 Behaviourism

First, let us compare mental fictionalism to analytic behaviourism. Like fictionalism, behaviourism also claims that ordinary talk about the mind is not committed to the existence of beliefs and desires as inner states. However, the behaviourist tries to avoid commitment to such internal states by giving a reductive analysis of mental talk in terms of behaviour. The key problem for behaviourism, of course, is that filling out such analyses proves to be impossible. There is no straightforward,

Mind as Metaphor: A Defence of Mental Fictionalism. Adam Toon, Oxford University Press. © Adam Toon 2023.
DOI: 10.1093/oso/9780198879626.003.0003

34 MIND AS METAPHOR

one-to-one correspondence between individual mental states and behavioural dispositions. Instead, it seems that any given mental state can involve any number of different behaviours, depending on the context and the person's other mental states. Recall our example from Chapter 1. Suppose we say that Mark believes that the No. 73 bus goes to Oxford Street. What behaviour might be associated with this belief? The possibilities seem endless: Mark might raise his hand (because the No. 73 is approaching and he wants to go to Oxford Street), nod his head (because someone at the bus stop asks him whether the No. 73 goes to Oxford Street), write '73' (because he friend texts him to ask which bus they should catch to meet him on Oxford Street), run down the street (because he's just seen the No. 73 approaching the bus stop) and so on. Given this, the prospects for translating talk about the mind into talk about behaviour seem pretty bleak.

Along with the behaviourist, the fictionalist also argues that folk psychology is not committed to the existence of beliefs and desires as inner states. The key difference between fictionalism and behaviourism, however, is that the fictionalist does not try to avoid ontological commitment by giving a reductive analysis of mental state attributions in terms of behaviour. The fictionalist does not try to translate talk about the mind into talk about behaviour. According to the analysis I introduced in Chapter 1, an utterance like 'Mark believes the No. 73 bus goes to Oxford Street' *means* roughly what the Cartesian takes it to mean: it means that there is a representation inside Mark's head with a particular content and certain characteristic causes and effects. The reason that we are not committed to the existence of such entities, according to the fictionalist, is simply because our utterance is not a straightforward assertion; instead, it is an act of pretence. Similarly, if we say, 'Crotone is on the arch of the Italian boot' our utterance means the same as if we were (oddly enough) to claim that there really *was* an enormous boot floating in the Mediterranean. Within the context of the game, however, our utterance is recognized as pretence (Walton 1990, 1993; Yablo 1998).

Our pretence involving the Italian boot allows us to make a genuine assertion regarding the location of Crotone. Similarly, our pretence regarding Mark's inner states allows us to make a genuine assertion regarding Mark and his behaviour. There is an important difference between the two cases, however. In the case of the Italian boot, the use of make-believe is dispensable. We can offer an alternative, literal paraphrase that captures the assertion we make about Crotone: we are claiming that Crotone lies on the southern coast of Italy, somewhere roughly between Capo Colonna and Taranto. Not all metaphors can be given a literal paraphrase, however. Stephen Yablo (1998, 250) notes the possibility that

> the language might have no more to offer in the way of a unifying principle for the worlds in a given content than that *they* are the ones making the relevant sentence fictional. It seems at least an open question, for example, whether the clouds we call *angry* are the ones that are literally *F*, for any *F* other than 'such

that it would be natural and proper to regard them as angry if one were going to attribute emotions to clouds'.

Yablo calls metaphors that resist literal paraphrase *representationally essential* (1998, 250; see also Walton 1993).

For the fictionalist, the failure of the behaviourist project suggests that the metaphors of folk psychology are representationally essential. If we say Mark believes the No. 73 bus goes to Oxford Street, we are saying that he is in some state S. Which state? Well, the state such that it is appropriate to pretend in the way that we do. There might be no alternative way of picking out this state, apart from by means of our pretence. In particular, there might be no way of specifying this state in terms of behaviour. Recall some of the vast range of different behaviours that might be associated with Mark's belief about the No. 73 bus: raising his hand as the bus is approaching, nodding his head when asked if it goes to Oxford Street, running down the road when it approaches the bus stop, and so on. Just as there might be no way of capturing what is common to all clouds that we call 'angry' (apart from that they each make it fictional that they are angry), so there might be no way of capturing what is common to each of these forms of behaviour (apart from that they each make it fictional that Mark has an inner representation with the content 'the No. 73 bus goes to Oxford Street'). The problems faced by behaviourism thus turn out to be an instance of a more general phenomenon, namely the difficulty of giving a literal paraphrase for metaphors and figurative language.

Many philosophers have worried that the metaphors that we use to talk about the mind might lead us astray (see especially Melser 2004, chs 8–10). For the fictionalist, however, the failure of behaviourism also serves to remind us of the positive value of metaphors. One of the main reasons that we value metaphors is that they expand our powers of expression, allowing us to say things that we could not say without them (Walton 1993, 2000; Yablo 1998). The same lesson applies to the metaphors that we use to describe the mind. These metaphors also add to our powers of expression, allowing us to assert things about people that we could not say otherwise. For this reason, it would be a mistake to demand that we spell out the content of these assertions, without recourse to metaphor (cf. Ross 2022). Philosophers have often pointed out that it is easier to give an intentional description of someone than a purely behavioural one (e.g. Wittgenstein 1953). Imagine, for example, trying to describe exactly what it was about the movements of the muscles in someone's face that led you to say that they were feeling happy. The same is true for many metaphors. Imagine trying to describe exactly what it was about someone's behaviour—precisely what they did or said, what tone of voice they used, and so on—that led you to say that they have a chip on their shoulder. The claim that our talk about the mind is metaphorical does not mean that such talk is somehow lacking or must play second fiddle to talk about behaviour.

36 MIND AS METAPHOR

2.1.2 Ryle

It is also helpful to compare fictionalism to Ryle's approach. Ryle's view is important in its own right, of course. It has also provided inspiration for more recent work, notably Eric Schwitzgebel's dispositionalist account of belief (2002). Ryle is often taken to be an analytic behaviourist. And yet in many respects this characterization is misleading, not least because Ryle never attempts to reduce talk about the mind to talk about behaviour. In fact, in his later work, Ryle describes both behaviourism and Cartesianism as 'category-howlers'—although he admits that he might be taken to have 'one leg and one ear' in the behaviourist camp (1979, 31). In Ryle's characterization, the behaviourist is a *Reductionist* whose slogan is 'Nothing But...' (1979, 79). She might insist, for example, that thinking is nothing but talking to ourselves. By contrast, Ryle tells us, the Cartesian is a *Duplicationist*. Her motto is 'Something Else as Well...' (1979, 79). Against the behaviourist, the Cartesian quite rightly points out that thinking cannot merely be talking to ourselves, since the very same sequence of words might be uttered intelligently or merely parroted. But the Cartesian wrongly concludes from this fact that the difference between the two—between talking intelligently and merely parroting something—must lie in some hidden ghostly utterance that has remained hitherto unseen.

The solution, Ryle insists, is to avoid both Reductionism and Duplicationism. Talk about the mind is neither to be reduced to talk about behaviour nor taken to describe some inner mechanism. Instead, the philosopher's task is to describe our ordinary talk about the mind as she finds it—to chart its 'logical geography' (1949, 17)—without mistakenly trying to shoehorn it into the mould of some entirely different domain of discourse, such as talk about the movements of muscles or machines. Once we begin to chart the landscape of folk talk about the mind, Ryle claims, we find that much of it is dispositional in character. To say that someone has a particular belief or desire is not to say that they have a particular cog in their inner machinery (either mechanical or para-mechanical), but to say what they are likely to do under certain circumstances. In this respect, Ryle's view is reminiscent of behaviourism. And yet, since he rejects the behaviourist's reductive project, Ryle is not faced with the impossible task of coming up with a definition that encompasses all the different behaviours that are associated with a given mental state. Instead, he can remain content with indicating the behaviour that we ordinarily associate with a particular mental state in different contexts. In doing so, Ryle can also happily refer to other mental states, or even to apparently private dispositions, such as our tendency to experience certain feelings (Ryle 1949; see also Schwitzgebel 2002).

How does fictionalism relate to Ryle's view? The key difference is one that we have encountered already: while Ryle takes the 'inner grotto' picture of the mind to be a category mistake imposed upon ordinary language, the fictionalist sees it as an important part of that language. The fictionalist can agree with Ryle that philosophers should aim to chart the 'logical geography' of talk about the mind. But she claims that its landscape takes a particular form—namely, a metaphorical one, with its central terms migrating from one area to another. One advantage of this approach is that it allows us to see how different aspects of someone's behaviour hang together. Consider once again the enormous variety of different behaviours related to Mark's belief: when he sees a No. 73 bus coming down the street, he sticks out his hand; when a friend wants to go to Soho he says 'get the No. 73 and walk down Wardour Street'; when the driver announces the bus will terminate at Tottenham Court Road, he jumps up from his seat in surprise; and so on. What on earth, we might wonder, do these different sorts of behaviour have in common? Why should our language count all of *these* as situations in which we should attribute this belief to Mark? Attending to the metaphorical structure of our talk about belief offers an answer to this question. Put simply, they are all respects in which Mark behaves *as if* he had 'No. 73 bus goes to Oxford Street' written down in an inner notebook that guides his actions.

Here we encounter another important feature of metaphors. As well as adding to our powers of expression, metaphors also structure our talk in certain domains (Lakoff and Johnson 1980). Consider the metaphor that treats argument as war. Suppose that we were to attempt to map the logical geography of our talk about arguments. We observe people engaging in arguments and describe in painstaking detail the language they use to talk about what they are doing. In a sense, our map might be perfectly accurate. And yet, as long as we remain ignorant of the metaphor that underpins this discourse, its structure would remain opaque to us. Why do people describe all these different conservational moves as *winning* and these as *losing*? Why does the author say she is offering a *defence* of her *position*? Why do her interlocutors seem so keen to identify parts of her discussion as *weak points* and conclude that what she says is *indefensible*? It is only once we recognize the guiding metaphor of argument as war that we see how such talk hangs together. For the fictionalist, the same is true of our ordinary talk about the mind. In a sense, our map of folk psychology might be perfectly accurate in charting the myriad of different behaviours that our mental language picks out. And yet it is only once we recognize the guiding metaphor of mind as an inner world that the terrain begins to make sense.

Consider a feature of our ordinary talk about the mind that we emphasized in Chapter 1. We take our thoughts to be about the world: they have contents that can be true or false. For instance, Mark's belief is about the No. 73 bus and Oxford Street, and what it says about them is true. On the face of it, this feature of our ordinary talk about beliefs seems at odds with the claim that they are simply dispositions to behave in certain ways, or to experience certain feelings. To put the point in Ryle's terms, it seems to be a category mistake to say that dispositions like these can be true or false. Of course, the things that people *say* can be true or false.

38 MIND AS METAPHOR

But this is not quite the same thing: we know that people don't always say what they believe, or believe what they say. Fictionalism offers a way to explain this. The fictionalist can agree that the aim of talk about beliefs is to identify patterns in someone's behaviour, or even the feelings they experience (we will discuss this further in Section 2.3). Fictionalism insists, however, that these patterns are picked out via metaphors that look to public representations, especially language. That is why we describe beliefs as having contents that can be true or false: to put it crudely, we talk as if beliefs have these features because language really does have them. (We will consider fictionalism's approach to intentionality in detail in Chapter 3.)

Fictionalism also differs from Ryle on the ontology of mental states. As we have seen, Ryle rejects both Reductionism and Duplicationism. He is content to chart mental language as it is, without trying to reduce it to talk about behaviour or the movements of an inner machine. As a result, his approach does nothing to advance the cause of materialism, nor does it aim to do so (see also Schwitzgebel 2002). Indeed, in *The Concept of Mind*, Ryle explicitly rejects materialism, as well as dualism and idealism (1949, 24). Fictionalism takes a rather different approach. Like Ryle, the fictionalist also tries to steer a middle course between Cartesianism and behaviourism. According to the fictionalist, Cartesianism's error is that it takes a metaphor too seriously: the inner world does not exist; it is merely a useful fiction. In contrast, behaviourism doesn't take this metaphor seriously enough: it overlooks the expressive power that it adds to our language. The upshot is that, although the fictionalist agrees with Ryle that talk about the mind cannot be reduced to talk about behaviour, she is still a 'Reductionist' in Ryle's terms. Strictly speaking, mind is 'nothing but' behaviour. It is simply that, when we come to describe this behaviour, we find we must invoke the metaphor of the mind as an inner world.

2.1.3 Instrumentalism

For the sake of argument, let us suppose that we could capture the principles underlying our ordinary belief and desire attributions in a theory. Following Dennett (1971), let us call this *intentional systems theory*. According to David Braddon-Mitchell and Frank Jackson (2007, 159), *instrumentalism* is the view that

> the predictive role of beliefs and desires is the whole story about them. There is nothing more to being a believer and a desirer, a thing with beliefs and desires, than being a being whose behaviour is well predicted by the principles of [intentional systems theory] ... All there is to say about belief and desire is encapsulated in the story about when ascriptions of belief and desire are true:

CONTRASTS, CAUSES, AND CONSCIOUSNESS   39

"*S* believes *P* and desires *Q*" is true iff (a) the behaviour of *S* is well predicted by intentional systems theory, and (b) the best belief and desire hypotheses according to the principles of intentional systems theory for predicting S's behaviour are that *S* believes *P* and desires *Q*.

Like behaviourism, instrumentalism rejects the idea that folk psychology is an attempt to describe our inner states. According to the instrumentalist, if I say that Mark believes the No. 73 goes to Oxford Street, I am not saying that he has any state inside his head with this content. I am saying that attributing this belief to Mark, together with many other mental states (e.g. the belief that buses take passengers, the desire to go to Oxford Street) yields a good prediction of his behaviour (e.g. jumping on the bus). Since it acknowledges that Mark's behaviour depends not upon any individual mental state, but a whole host of them taken together, instrumentalism avoids the key difficulty facing behaviourism, of trying to find a one-to-one mapping between individual mental states and behavioural dispositions.

An important challenge for instrumentalism is provided by the Blockhead thought experiment (Block 1981). Blockhead is a creature whose outward behaviour matches our own, but whose inner machinery consists of a giant 'look-up' table determining its response to each possible input. Suppose that Blockhead is having a conversation with Jane, who happens to remark 'What a nice day!' On receiving this input, Blockhead promptly looks up Jane's remark in a vast list containing all the possible remarks that she might have made (e.g. 'What a boring day!', 'What a nice dog!' etc.). Finding the correct entry, it then issues a pre-programmed response ('Yes, isn't it?' AND *smile*). However impressive Blockhead's outward behaviour might seem, once we take a look inside and see how it works we might be somewhat disappointed. As Block puts it, 'the machine has the intelligence of a toaster' (1981, 21). Intuitively, it is argued, we would not count Blockhead as a genuine thinker with fully fledged beliefs and desires. And yet it seems that instrumentalism cannot account for this intuition. After all, according to the instrumentalist, all we mean when we say that someone has beliefs or desires is that their behaviour can be predicted by intentional systems theory—and this is perfectly true of Blockhead.

For its proponents, the Blockhead thought experiment shows that folk psychology cares about more than outward behaviour. As Braddon-Mitchell and Jackson put it, 'it matters how the trick is turned inside' (2007, 163). Set against cases like Blockhead, however, those more sympathetic to instrumentalism have offered their own thought experiments intended to elicit precisely the opposite intuition. Dennett, for example, asks us to

[s]uppose, for the sake of drama, that it turns out that the sub-personal cognitive psychology of some people turns out to be dramatically different

40 MIND AS METAPHOR

from that of others. One can imagine the newspaper headlines: 'Scientists Prove Most Left-handers Incapable of Belief' or 'Startling Discovery—Diabetics Have No Desires.' But this is not what we would say, no matter how the science turns out.

And our reluctance would not be just conceptual conservativism, but the recognition of an obvious empirical fact. For let left- and right-handers (or men and women or any other subsets of people) be as internally different as you like, we already know that there are reliable, robust patterns in which all behaviourally normal people participate—the patterns we traditionally describe in terms of belief and desire and the other terms of folk psychology.   (1987, 235)

Our intuitions would therefore seem to be somewhat conflicted. On the one hand, cases like Blockhead seem to show that it is not only behaviour that counts: our ordinary concepts of belief and desire make reference to our inner workings. On the other hand, it is hard to imagine that we would give up talking about belief and desire, whatever future cognitive science discovers about what is going on inside our heads.

Fictionalism allows us to explain these conflicting intuitions. The key point to notice is that, unlike instrumentalism, fictionalism does not claim that all we mean when we attribute mental states is that these attributions work. In fact, as we have seen already, the fictionalist takes an utterance like 'Mark believes the No. 73 bus goes to Oxford Street' to *mean* what the Cartesian does: taken literally, it means that there is a particular sort of internal state inside Mark's head. It is by pretending to assert this that we make a genuine claim about Mark's behaviour. For the fictionalist, then, our ordinary concepts of belief and desire are not solely concerned with behaviour; instead, they do indeed make reference to our inner workings. It is simply that this reference is metaphorical, not literal. As a result, fictionalism has the conceptual resources to explain our response to cases like Blockhead: it is perfectly true to say that Blockhead lacks the internal machinery that is invoked by our ordinary concept of mind. At the same time, fictionalism also explains our response to Dennett's scenario: even if the folk vision of our inner workings proves to be wrong, we would continue to use it, since its real purpose is to make claims about behaviour.

Once again, this point can be put in terms of a general feature of metaphors. We often distinguish between metaphors that are 'alive' and those that are 'dead'. Consider the term 'blockbuster' (Gentner and Bowdle 2008, 118; Hall 2014). In its original use, arising in the Second World War, 'blockbuster' referred to a bomb large enough to destroy an entire block of a city. Shortly afterwards, the term was used metaphorically: we began to talk about films (and novels, plays, etc.) *as if* they had the same shattering impact. Nowadays, most people are unaware of the term's original sense. It seems that 'blockbuster' is a dead metaphor. Its original, literal meaning no longer plays a role in its current usage.

Instead, the term has simply acquired a new literal sense—meaning, roughly, a film (or novel, play, etc.) that is destined for, and finds, a mass audience. Is the same true of ordinary talk about the mind? Fictionalism claims that such talk is guided by the metaphor of an inner world. And yet it might be argued that, even if this were true at some point in time, the metaphor is now dead. Terms like 'belief' and 'desire' no longer look to inner sentences or pictures to gain their meaning, just as 'blockbuster' no longer looks to 1,000-pound bombs. Instead, they have simply acquired a new literal sense that applies to certain forms of behaviour. If we take this line, then it seems that fictionalism would collapse into instrumentalism: we would lose fictionalism's distinctive claim that we that talk about people's behaviour *via* the pretence of an inner world. Our response to thought experiments like Blockhead shows that this would be a mistake, however. For what Blockhead shows is that our ordinary concept of mind still invokes the idea of an inner world. The metaphor is alive and kicking.

2.1.4 Dennett

Daniel Dennett is often taken to be the leading proponent of instrumentalism. Dennett himself has objected to this label, however (1987, 1991b). His own approach is more nuanced and Dennett's claims about the nature of mental states are often at odds with a straightforward or 'naïve' instrumentalism like that explored in Section 2.1.3. Dennett has also objected to being described as a fictionalist (1987, 1991b, 2022; see also Appiah 2017 and Toon 2019). As we shall see, there are certainly important differences between Dennett's view and mental fictionalism in the form in which I defend it in this book. This will emerge most clearly in Chapter 3, when we consider fictionalism's approach to intentionality. In spite of this, however, fictionalism does share some important similarities with Dennett's approach. In this section, my aim will be to show that a fictionalist analysis of folk psychology can make sense of aspects of Dennett's view that his critics have found most problematic.

At the heart of Dennett's view are three interpretive strategies or *stances*. The *physical stance* is the strategy adopted in the physical sciences. When we adopt this stance, we endeavour to explain and predict an object's behaviours using known physical laws. For example, I predict that the apple I am holding will fall if I drop it because I know it has mass and that it is subject to the law of gravitation. The *design stance* is the strategy we often use to explain and predict the behaviour of artefacts. When we adopt this stance, we assume that an object was designed to perform a particular function. For example, I predict that hitting *these* keys on my computer's keyboard will cause *this* text to appear on the screen, since I assume it was designed to do this. Notice that I can make this prediction, even if I know nothing about how my computer achieves this feat. Of course, if the computer

42 MIND AS METAPHOR

breaks, then this knowledge might become important. In Dennett's terms, I might have to resort to the physical stance (e.g. I realize that I have spilled tea on my keyboard, which has damaged an electrical circuit).

The final, and for our purposes most important, interpretative strategy is what Dennett calls the *intentional stance*. This is the strategy that we adopt when it comes to making sense of people, either in everyday life or in the sciences (e.g. psychology, sociology, etc.). When we adopt the intentional stance, we treat someone as a rational agent with beliefs and desires, who will act on the basis of those beliefs and desires. For example, I predict that Mark will jump on the No. 73 bus because I think that Mark believes that it goes to Oxford Street and wants to go to Oxford Street, and that he will act accordingly. As Dennett points out, we also adopt the intentional stance when dealing with animals (e.g. we predict the dog will come inside because it wants food and believes its food bowl is in the kitchen) and even some artefacts (e.g. we predict the chess computer will move its queen because it wants to win and believes that this is the best move). The theoretical perspective that underpins the intentional stance is one that we have encountered already in Section 2.1.3, namely what Dennett calls *intentional systems theory* (1971). As we saw, intentional systems theory aims to capture the principles that guide our ordinary attributions of belief and desire. It is an attempt to give an explicit formulation of folk psychology as a theory—albeit, perhaps, suitably regimented and tidied up here and there (Dennett 1987).

Dennett is keen to stress that, just as with the design stance, we can adopt the intentional stance towards some system, even if we know nothing about its internal workings. For example, I can predict that the chess computer will move its queen, even if I know nothing about how exactly it is programmed or even how computers work. Of course, again as with the design stance, predictions made from the intentional stance can prove to be wrong, if the system fails to meet our assumption that it is rational. In such cases, we might have to fall back on the design stance to explain its behaviour (e.g. it fails to move its queen, since it is designed to let beginners win) or even the physical stance (e.g. it fails to move its queen, since I've knocked my tea over the computer as well as the keyboard). The same lesson applies, Dennett suggests, when we adopt the intentional stance towards people: '[t]he central epistemological claim of intentional systems theory is that when we treat *each other* as intentional systems ... we are similarly finessing our ignorance of the details of the processes going on in each other's skulls' (Dennett 2009, 341).

In light of this, the key question becomes: what are we doing when we adopt the intentional stance towards someone (or alternatively, towards an animal or artefact)? And what does this tell us about the nature of mental states? One option, it would seem, is to turn to instrumentalism. Dennett explicitly describes folk psychology as 'instrumentalistic' (e.g. 1987, 52) and some of his remarks certainly seem to fit well with instrumentalism. For example, Dennett writes that 'any object...whose behaviour is well predicted by [the intentional stance] is in the fullest sense of the word a believer. *What it is* to be a true believer is to be an *intentional system*, a system whose behaviour is reliably and voluminously predictable via the intentional stance' (1987, 15; emphasis in original). Or again, Dennett says that '*what it means* to say that someone believes that *p*, is that the person is disposed to behave in certain ways under certain conditions' (1987, 50; emphasis in original). These comments seem to express a straightforward instrumentalist position like that discussed in Section 2.1.3.

Elsewhere, however, Dennett's writing reveals a more nuanced picture of folk psychology that is less easy to characterize. For example, in an influential discussion, Dennett asks:

> What are beliefs? Very roughly, folk psychology has it that *beliefs* are information-bearing states of people that arise from perceptions and that, together with appropriately related *desires*, lead to intelligent *action* (1987, 46; emphasis in original).

Critics have seen a tension in his view here: how can our ordinary notion of belief signify *both* an information-bearing state *and* a behavioural disposition (Bennett and Hacker, 2003, 415)? For his part, however, Dennett suggests that folk psychology *itself* is somewhat conflicted. He draws on a distinction due to Hans Reichenbach (1938), between *illata* and *abstracta*. *Illata* are entities posited by a theory, like electrons or genes. In contrast, *abstracta* are mere logical constructs that the theory employs, like centres of gravity. In general, Dennett claims, folk psychology treats beliefs and desires as abstracta rather than illata. And yet he acknowledges that the true picture might be somewhat more complicated, and even conflicted. In fact, Dennett concedes, '[t]he *ordinary* notion of belief no doubt does place beliefs somewhere midway between being *illata* and being *abstracta*' (1987, 55; emphasis in original).

We have already seen that Dennett rejects the label 'fictionalist'. And yet, I suggest, mental fictionalism—at least, as I understand the view—can make sense of the apparent conflict that Dennett identifies in folk psychology. The critic asks how our ordinary notion of belief can signify both an information-bearing state and a behavioural disposition. Notice, however, that this is precisely what the fictionalist would expect to be the case. Fictionalism says that we pretend that people have information-bearing states in order to make assertions about their behavioural dispositions. It is no surprise, therefore, if our folk concept of belief incorporates both of these notions. Like all metaphors, our ordinary concept of belief has a dual character: it looks both to its source (i.e. information-bearing states, such as written language) and its target (i.e. behavioural dispositions) at the same time. This is the characteristic feature of metaphors: they bring together two different, often disparate, domains in order to work their magic.

44 MIND AS METAPHOR

A fictionalist analysis can also make sense of other well-known claims that Dennett makes about folk psychology. Consider the question of whether attributions of mental states are true or false. Dennett (1987, 72) distances himself from fictionalism, which he takes to be the claim that theoretical statements are useful falsehoods. And he also denies that talk about mental states is neither true nor false. Instead, Dennett argues, talk about beliefs and desires can be true, although this is a truth which 'one must understand *with a grain of salt*' (1987, 72–3; emphasis in original). To explain this idea, Dennett draws a parallel with centres of gravity:

> I should have forsworn the term [instrumentalism] and just said something like this: My *ism* is whatever *ism* serious realists adopt with regard to centres of gravity and the like, since I think beliefs... are *like that*—in being *abstracta* rather than part of the 'furniture of the physical world' and in being attributed in statements that are *true* only if we exempt them from a certain familiar standard of literality. (1987, 72)

Thus, Dennett suggests, claims about beliefs and claims about centres of gravity can both count as true, even if neither beliefs nor centres of gravity exist. Fictionalism offers a way to make sense of this idea (cf. Hutto 2013). When we say that a system has particular centre of gravity, we pretend that its mass is located at a certain point. Taken literally, this might be false. And yet, when we attribute a centre of gravity to a system, we also make a genuine assertion: we claim that the system's mass is distributed in a certain manner M, such that it is appropriate to pretend that all the mass is located at a particular point. And this claim can be straightforwardly true. Similarly, according to the fictionalist, when we claim that someone has a certain belief, we pretend that they have a particular inner state. Taken literally, this might be false. And yet, when we attribute a belief, we also make a genuine assertion: we claim that they are in a particular state S, such that is appropriate to pretend in this way. And this claim can be straightforwardly true.

Dennett also famously insists that, although beliefs and desires might not exist, the patterns picked out by the intentional stance are nevertheless entirely real, objective features of the world (Dennett 1987, 1991b). Moreover, Dennett argues, these patterns would be overlooked by Martians who knew all there was to know about the physical world, but failed to adopt the intentional stance. Once again, I think that a fictionalist analysis can help us to understand how this might be the case. Suppose that we say 'the clouds over Exeter were angry all day'. There are no (literally) angry clouds. And yet when we say this we might capture an entirely real, objective feature of the world: the fact that the clouds over Exeter on a particular day were a certain way, such that our pretence is appropriate. And if describing clouds as angry is a metaphor that resists literal paraphrase—so that we

CONTRASTS, CAUSES, AND CONSCIOUSNESS 45

are unable to give a description in the language of the physical world that specifies what it is that all clouds that we would fictionally count as angry have in common—then the pattern exhibited by clouds over Exeter today might be inaccessible to someone who doesn't recognize this particular game. Similarly, even if beliefs and desires do not exist as discrete inner states, pretending that they do might allow us to pick out perfectly real and objective features of the world. And if talk of beliefs and desires is a metaphor that resists literal paraphrase—so that we are unable to give a description in the language of the physical world that specifies what it is that all people possessing a particular belief or desire have in common—then these patterns might be overlooked by beings that did not take part in the game of folk psychology.

So it seems that fictionalism can explain some of Dennett's influential, but apparently problematic, claims about the nature of folk psychology. Dennett's view has also received trenchant criticism, of course. In many cases, I suggest, fictionalism has the resources to respond to these criticisms. For instance, Bennett and Hacker object that

> [i]n order to be able to treat a being *as if* it believed that *p* or wanted to *V*, we must know what it *is* to believe or want something... But on Dennett's account the adoption of the intentional strategy towards an intentional system, including human beings, is *never* other than an "as if". But if so, there is no actuality in terms of which we might intelligibly cash the pretence... (2003, 426)

Consider an ordinary case of pretending, such as a child pretending to be a bear. In order to be able to pretend to be a bear, it seems, the child must know what a bear is. And yet Dennett insists that *all* attributions of belief and desire involve 'as if' thinking, or pretence. How can we pretend that people have beliefs and desires if we do not know what it is actually to have a belief and desire? What is the content of our pretence? (For similar objection to mental fictionalism, see Bourne and Caddick Bourne 2020, 178.)

Whether or not this is a fair criticism of Dennett's position, I think it is clear that the fictionalist can respond to this worry. It is true that, for the fictionalist, all attributions of beliefs and desires (whether to people, animals, or artefacts) are to be understood in terms of pretence. There is no difficulty in filling out the content of this pretence, however. For, according to the fictionalist, our pretence is filled out by looking to public representations, especially spoken and written language. When we attribute a belief or desire to someone, we pretend that they have an inner representation that expresses the content of their belief or desire. And we know what it is actually to have such a representation, since we know what it is to say sentences out loud and write them down on paper. We also know what it is to record information about the world around us using pen and paper and, later on, to act upon that information. In these and many other ways, the fictionalist argues,

46   MIND AS METAPHOR

it is the world of public representations that gives content to our pretence. It is public representations that show us what it is (literally) to have content, to have meaning, to be about the world and to be capable of being true or false. It is these properties that we project, metaphorically, onto the mind as an inner world.

It is here, however, that we encounter a fundamental difference between Dennett's position and mental fictionalism. For, unlike Dennett, the fictionalist claims that, in many respects, the intentionality of public representations comes before that of mental states. We shall examine fictionalism's approach to intentionality in Chapter 3, where its departure from Dennett's view will become clearer. As we shall see, this aspect of mental fictionalism allows it to avoid an important objection levelled against Dennett's intentional stance, namely that it suffers from *cognitive collapse*.

2.1.5 Prefix-fictionalism

There are different sorts of fictionalism. One prominent approach is *prefix-fictionalism* (Armour-Garb and Woodbridge 2015). According to this approach, statements made in some area of discourse should be understood as having a tacit prefix that reads 'In the fiction...'. For example, a prefix-fictionalist about mathematics argues that a statement like '2 + 2 = 4' should be understood as elliptical for the prefixed claim, 'In the mathematical fiction, 2 + 2 = 4'. The guiding analogy here, of course, is with statements we might make while reading a work of fiction. Suppose that, while reading *The Hound of the Baskervilles*, we happen to remark that 'Holmes smokes a pipe'. According to prefix-fictionalism, our statement is short for 'In the *Hound of the Baskervilles*, Holmes smokes a pipe'. By understanding our utterances in this way, it is argued, we avoid the apparent commitment to problematic entities, such as mathematical objects or non-existent detectives (though whether we do in fact avoid commitment in this way is debateable; see Friend 2007). Applied to folk psychology, prefix-fictionalism would claim that a statement like 'Mark believes the No. 73 goes to Oxford Street' should be taken to be elliptical for the prefixed claim, 'In the folk psychological fiction, Mark believes the No. 73 goes to Oxford Street' (Parent 2013; Wallace 2022). This is not the sort of fictionalism that I have proposed in this book. Instead, the view I have defended is a form of *pretence-fictionalism* (Armour-Garb and Woodbridge 2015): I have suggested that talk about mental states should be understood as an act of pretence, rather than involving an implicit prefix. Is there a reason why we should prefer the pretence form of mental fictionalism?

The prefix form of mental fictionalism is a version of what Yablo (2001) calls *meta-fictionalism*: according to the prefix-fictionalist, utterances that appear to be about people's mental states turn out to be claims about the contents of a particular theory, namely folk psychology (see also Eklund 2019). Yablo

CONTRASTS, CAUSES, AND CONSCIOUSNESS   47

(2001, 75–6) presents a number of challenges for meta-fictionalism, focusing primarily on mathematical fictionalism. First, Yablo argues, meta-fictionalism has problems dealing with modal claims. We regard claims like '2 + 2 = 4' as necessarily true. According to the prefix-fictionalist, however, when we say '2 + 2 = 4' we are claiming that 'In standard mathematics, 2 + 2 = 4'. And yet, arguably, this claim is not necessarily true, since standard mathematics might perhaps have turned out differently. Second, Yablo argues that meta-fictionalism faces 'problems of concern'. Suppose that we say, 'the number of people starving is large and rising'. When we say this, what we care about is people and their desperate situation. And yet, according to the prefix-fictionalist, what we are really talking about is the content of standard mathematics. Third, Yablo raises a related, phenomenological worry. When we say, 'the number of people starving is large and rising', we do not feel remotely as if we are talking about the content of standard mathematics; instead, we feel that we are talking about people.

Each of these difficulties confronts prefix versions of mental fictionalism. First, consider modality (cf. Wallace 2022, 38). Of course, unlike a claim like '2 + 2 = 4', we don't take attributions such as 'Mark believes the No. 73 goes to Oxford Street' to be necessarily true. And yet prefix fictionalism still seems to run into difficulties here. Consider a possible world in which folk psychology turned out differently (perhaps our visionary Jones never lived, or took a different model as the inspiration for his revolutionary account of people's behaviour). According to prefix-fictionalism, in such a scenario, a claim such as 'Mark believes the No. 73 goes to Oxford Street' would be false, even if all the facts about Mark and his bus-related behaviour were exactly the same. This seems wrong: intuitively, I think, we should still count this as a world in which Mark believes the No. 73 goes to Oxford Street. Or, to take another example, suppose that we grant that, since Freud's work, the notion of unconscious desires has become part and parcel of folk psychology. Again, it seems wrong to say that, had Freud never written, then all our attributions of unconscious desires would be false. Second, consider the problem of concern. In general, when we talk about people's mental lives (rather than, say, about psychology or philosophy of mind), what we care about are people and their behaviour, not the contents of folk psychological theory. Finally, when we say, 'Mark wants to go shopping' or 'Mark wants a new job', it certainly feels as if we are talking about Mark, not about the way the folk talk about Mark.

The pretence version of mental fictionalism avoids each of these worries. According to the analysis I proposed in Chapter 1, our attributions of mental states are not claims about the contents of folk psychological theory; instead, they are acts of pretence that serve to make claims about people and their behaviour. This allows the pretence fictionalist to respond to each of the difficulties facing prefix fictionalism. If we say (correctly) that 'Mark believes the No. 73 goes to Oxford Street', the claim we make is true in virtue of facts concerning Mark and

48   MIND AS METAPHOR

his behaviour. If the contents of folk psychology were to change while the facts about Mark remained the same, then our claim would still be true. Similarly, our attributions of unconscious desires would retain their present truth values, even if Freud had never written. And the pretence fictionalist has no problem explaining why it is that, when we attribute mental states to Mark, it is Mark himself that we care about and whom we feel we are talking about: according to the pretence analysis, that is precisely what we are doing, although we do so via the use of pretence.

The pretence version of mental fictionalism also avoids a further difficulty confronting prefix-fictionalism. As we have seen, the prefix-fictionalist is guided by an analogy between folk psychology and works of fiction. And yet there seems to be an important difference between the two cases. It is clear which fiction underpins our talk about Sherlock Holmes: we can point to our copy of *The Hound of the Baskervilles*. There is no text that sets out the principles of folk psychology. Indeed, such principles are notoriously difficult to formulate explicitly. In this respect, mental fictionalism might seem to be in a worse position than fictionalism in other domains. For instance, a mathematical fictionalist could direct our attention to the nearest maths textbook for its guiding fiction. Of course, there are ways that the prefix fictionalist might try to respond to this worry. For instance, perhaps she will insist that it is enough to regard folk psychology as an implicit theory, without actually trying to write it down (Wallace 2022, 31–2). For our purposes, the important point is that the problem does not arise for pretence fictionalism. All pretence fictionalism requires is that we have a reasonably coherent set of rule-governed practices for attributing mental states to people based on their behaviour. It does not require that we are able to distil any folk psychological principles or laws by reflecting on those practices, nor that we regard folk psychology as a theory. The rules of most make-believe games—even children's games with dolls—are complex and difficult to formulate explicitly.

2.2 Explaining Behaviour

Each of the anti-Cartesian views that we have considered in this chapter faces a common objection: they all seem to deny the obvious fact that mental states are causes. The same worry applies to mental fictionalism (Sprevak 2013). After all, surely fictions cannot be causes? Let us now turn to consider this issue.

2.2.1 Wires and Pulleys

We saw in Chapter 1 that our ordinary talk about the mind appears, at least at first glance, to treat mental states as causes. We seem to talk as if our thoughts can be

CONTRASTS, CAUSES, AND CONSCIOUSNESS   49

caused by, and cause, events in the world. Why did I reach for the kettle? Because I had a long meeting and it made me want a cup of tea. When we say this, it is natural to think that we are offering a causal explanation: the long meeting has caused me to want a cup of tea which, in turn, has caused me to reach for the kettle. In fact, at least two different sorts of causal connection seem to be involved here: from states of the world to states of the mind (the meeting causes the desire for tea) and vice versa (the desire for tea causes reaching for the kettle). We also seem to talk as if thoughts can cause other thoughts. Why do I think I'll drink lots of tea today? Because I think that I'm writing today and that if I'm writing, I'll drink lots of tea. Again, it seems natural to think that this is a causal explanation: my belief that I'm writing today, together with my belief that, if I'm writing, I'll drink lots of tea causes me to believe that I'll drink lots of tea today. How are we to make sense of these sorts of explanations of people's behaviour? Are they really causal explanations? If they are, how can we understand this sort of causation?

Cartesians claim that folk psychological explanations of people's behaviour are causal explanations: we explain people's behaviour by describing its inner causes. They are then faced with the task of understanding these causal links. Mental causation is a notorious problem for dualists, of course: if mind and body are made from radically different substances, how can they interact? At first glance, it looks less tricky for materialists: if the mind is part of the physical world, then interactions between mind and body do not seem especially mysterious. And yet, even for materialists, a number of puzzles remain. For our purposes, the most important concern the content of thoughts. Consider what John Haugeland calls *the paradox of mechanical reason* (1985, 39). Suppose that thoughts are inner symbols with syntax and semantics. How can one set of inner symbols (e.g. the symbols meaning *if I'm writing, I'll drink tea*, and *I am writing*) cause another set of inner symbols (e.g. the symbol meaning *I'll drink tea*)? If this is to count as genuine reasoning then it seems as if the process must somehow depend upon the meaning of these symbols. But how does this work? After all, the symbols themselves might be physical states (brain states, perhaps), but is their *meaning* something physical? Do we need to posit some inner homunculus who *under-stands* these symbols? If so, then our account seems circular: we have explained thinking by positing an inner thinker!

The computational theory of mind promises to solve this problem. As we noted in Chapter 1, some of the operations that we carry out on symbols seem to depend entirely on their syntax, not their semantics. From $p{\rightarrow}q$ and p we can infer q, whatever the contents of p and q. And yet our manipulation of these symbols also makes sense if we consider their meaning. (For example, if p is *I'm writing* and q is *I'll drink tea*.) Computers are able to carry out these operations purely mechan-ically, using symbols encoded electronically rather than by marks on paper. This is a remarkable innovation, since it seems to resolve the paradox of mechanical reason: it shows that symbols can be manipulated purely mechanically in a

50   MIND AS METAPHOR

manner that nevertheless respects their meaning. And yet we might still have reservations. For, strictly speaking, it seems that the *content* of the symbols remains causally inert. The computer is a syntactic machine: the presence of one symbol causes the presence of another solely due to its syntactic properties (e.g. because it has a certain voltage or current). And yet surely the content of our thoughts is relevant for explaining their effects. My belief that tea is refreshing causes me to reach for the tea caddy (at least in part) because it is *about* tea. If it had been about coffee instead, then I might have reached for a jar of coffee rather than the tea caddy.

The causal properties of thoughts thus prove difficult to understand, even for materialists. If we take a step back, we might also wonder whether our folk psychological explanations are causal explanations after all. We typically think of causal connections as a contingent matter. In the actual world, fire causes smoke—but it seems perfectly possible that it might have given off clouds of bubbles instead. The situation with mental states seems rather different. Consider the link between mind and behaviour. Surely it is not only a contingent matter that the desire for ice cream tends to lead to eating ice cream? Or that believing that the No. 73 goes to Oxford Street tends to lead to saying things like 'the No. 73 goes to Oxford Street'? Instead, these links seem to be conceptual ones. At least part of what we *mean* when we say that someone desires something is that, under certain circumstances, they will try to get it. This is part of our concept of desire. Similarly, at least part of what we mean when we say that someone believes *p* is that, under certain circumstances, they will say it. This is part of our concept of belief. All this makes it hard to see the link between mental states and behaviour as an entirely contingent or accidental one. We have no trouble imagining a world in which fire gives off bubbles, but can we really imagine a world in which people have a burning desire for ice cream but would never, under any circumstances whatsoever, try to eat it?

Considerations like these typically lead anti-Cartesians to interpret our folk psychological explanations rather differently to Cartesians. According to behaviourists, to attribute a belief or desire is to attribute a behavioural disposition: to say that someone wants ice cream is simply to say that, under certain circumstances, they will eat ice cream. This yields a rather different view of folk psychological explanation. 'Why is Mary taking that tub out of the freezer?' 'Because she wants some ice cream.' According to the behaviourist, our use of 'because' here does not serve to indicate a cause of Mary's behaviour. Our explanation is not a causal one. Instead, we are explaining this particular instance of Mary's behaviour by showing how it fits into a more general behavioural disposition that she exhibits. Ryle's view of folk explanation is similar. As he puts it, 'to explain an act as done from a certain motive is not analogous to saying that the glass broke because a stone hit it, but to the quite different type of statement that the glass broke, when the stone hit it, because it was fragile'

CONTRASTS, CAUSES, AND CONSCIOUSNESS   51

(1949, 84). Similarly, both instrumentalism and Dennett's intentional stance see folk psychological explanations as attempts to fit someone's behaviour into an overall pattern, rather than identify its inner causes.

Unsurprisingly, then, Cartesians and anti-Cartesians disagree over how we ought to interpret folk psychological explanations of behaviour. According to the Cartesian, folk explanations are causal explanations. As a result, they face the challenge of understanding mental causation. They must also find a way to explain the conceptual connections that exist between mental states and behaviour. In contrast, anti-Cartesians claim that folk explanations are not causal, but instead serve to fit someone's behaviour into a larger pattern. In doing so, they would seem to avoid the challenge of explaining mental causation, at least as it is normally understood. The main challenge for anti-Cartesians, it would seem, is to explain (or perhaps explain away) our intuitive sense that there *are* causal interactions between mental states and behaviour. For Ryle, to inquire after the inner causes of behaviour is to ask a mistaken 'wires and pulleys' question that is part and parcel of our dubious inheritance from Descartes. And yet others find it difficult to see how we could possibly give up this idea. As Fodor memorably puts it,

> if it isn't literally true that my wanting is causally responsible for my reaching, and my itching is causally responsible for my scratching, and my believing is causally responsible for my saying...then practically everything I believe about anything is false and it's the end of the world.   (1989, 77)

2.2.2 Imagined Mechanisms

Fictionalism offers a way to resolve these conflicting intuitions over folk psychological explanations of behaviour. Recall the central elements in the fictionalist analysis. According to the fictionalist, talk about the mind is metaphorical: we talk *as if* people had inner analogues of outward representations, such as notebooks, diaries, shopping lists, and itineraries. When we attribute mental states, we do not claim that people actually have such representations inside their heads that cause them to behave as they do. Instead, we merely pretend that this is the case. And yet, by means of our pretence, we do make a genuine assertion about someone: we claim that they are in a certain state—namely, the state such that our pretence is appropriate. What does this mean for folk psychological explanations of behaviour? How does fictionalism interpret these explanations?

Consider an example. Why did Mark say 'the No. 73 goes to Oxford Street' when the woman at the bus stop asked him about its route? Because he believes that the No. 73 goes to Oxford Street. What sort of explanation is this? Like other anti-Cartesians, the fictionalist will deny that it is an attempt to identify the

52 MIND AS METAPHOR

inner causes of Mark's behaviour: we are not claiming that there is an inner representation that causes Mark's utterance; we are simply pretending that this is the case. The fictionalist will also agree that the link between belief and behaviour is, in a sense, a conceptual one. Attributing a mental state to someone involves pretending not merely that they possess a certain inner representation, but also that it guides their behaviour in a particular way. We use notebooks or diaries to guide our behaviour in different ways to shopping lists or itineraries—and so the metaphors based on these practices differ in similar ways. The images invoked by these metaphors are, I think, thoroughly 'homuncular': if we imagine that memory is an inner notebook, we give little thought to the implication that someone must be reading it.

In many respects, then, fictionalism comes down firmly upon the side of the anti-Cartesians. And yet fictionalism does not see our hankering after inner mechanisms as merely a misguided product of Cartesianism. What Ryle disparagingly calls a 'wires and pulleys' conception of the mind is not idle or merely a philosopher's invention, for it is precisely by entertaining the idea of these imaginary inner mechanisms that we come to make sense of people. The fictionalist can also allow that, even if they don't describe inner causes, folk psychological explanations *are* genuine causal explanations. Suppose we say 'the angry clouds caused Ruth to fetch her umbrella'. When we say this, we are claiming that the clouds were in some state S—a state such that our pretence is appropriate—and this state S caused Ruth to fetch her umbrella. Even if there are no (literally) angry clouds, this is still a causal explanation. We have simply picked out the relevant state of the weather via a metaphor. Similarly, if we say Mark's belief caused him to reply 'the No. 73 goes to Oxford Street' then we are claiming that Mark is in some state S—a state such that our pretence is appropriate—and this state S caused him to say what he did. Even if Mark has no inner representation expressing his belief, this is still a causal explanation. We have simply picked out Mark's overall state via our folk psychological metaphor. (For a similar analysis, see Dennett 1987, 56–7.)

2.3 Self-knowledge

We can now consider another important challenge that is levelled against anti-Cartesian views in general, and mental fictionalism in particular. This challenge concerns our knowledge of our own mental states. So far, I have focused on our attribution of mental states to other people. I have said that we use metaphors to make sense of their behaviour. What are we doing when we attribute mental states to ourselves? Isn't this a rather different matter? After all, don't we know our own thoughts more intimately, and in a different manner, than we know the thoughts of others? The apparent asymmetry between our knowledge of ourselves and our knowledge of others is commonly taken to show that we possess a special way of finding out about our own mental states, known as *introspection*. The term is revealing, of course: it suggests that, just as we can look outwards to the external world of tables, chairs, or mountains, we can also look inwards to the internal world of our thoughts. Indeed, introspection is often taken to be more dependable than our ordinary senses: our knowledge of our own thoughts is said to be far more secure than our knowledge of tables, chairs, or mountains. If that is right, then how can the fictionalist doubt that such things exist? Can't we just look inside ourselves and *see* that they do?

2.3.1 Finding Out about Ourselves

The first point to remember is that we are concerned with thoughts. For other aspects of the mind, the situation might be somewhat different. In the case of sensations, for instance, there does seem to be a marked asymmetry between our knowledge of ourselves and our knowledge of other people. Consider pain. We often know that other people are in pain, of course. If a friend's knee is hurting, he might tell us so, or we might notice that he is limping, or see the pain etched in his face as he tries to stand up. Still, this seems very different from the way that we would find out that our own knee was hurting. In fact, it seems odd to say that we would find this out at all. We'd just feel it. In this manner, it seems that our knowledge (if we can even call it that) of our own sensations is different, and more immediate, than our knowledge of other people's sensations. It also seems less prone to error. Ordinarily, we might have no doubt that our friend was in pain. Still, we realize that it's at least conceivable that he might be trying to deceive us: perhaps he's looking for sympathy or trying to get out of going for a strenuous hike at the weekend. Things look rather different in our own case. Indeed, many would say that there is simply no possibility for error here: if we think we are in pain, we are in pain; conversely, if we are in pain, we know we are.

There is therefore an important asymmetry between our knowledge of our own sensations and our knowledge of other people's. Of course, it is one thing to recognize this asymmetry; is another to explain it or assess its import. Here there will be considerable disagreement. Ryle criticizes Cartesians for talking as if the asymmetry arises from the fact that sensations are strange, inner objects that, unlike ordinary objects like tables and chairs, can only ever be seen by one person (1949, Chapter VII). Although I am sympathetic to Ryle's view on the nature of sensations, we need not enter into these debates here. Our analysis concerns thoughts, not sensations. For our purposes, we can simply acknowledge that we do indeed have sensations, and that we learn about them in a different manner to that in which we learn about other people's. We can also accept that our knowledge of our own sensations is, in some sense, especially secure. For us, the key question is whether the same might be said of our thoughts. In fact, there are two

54   MIND AS METAPHOR

questions here. First, is there an important asymmetry between our knowledge of our own thoughts and our knowledge of other people's? Second, does this show that we have a special way of finding out about our own thoughts?

We should begin by reminding ourselves that, sometimes at least, other people are a better judge of our thoughts than we are. I might loudly declare my belief in the importance of looking after the environment, but if I drive a petrol-guzzling car and drop litter wherever I go, you are probably right to conclude that I don't really believe what I say. Often, though, we are best placed to know our own thoughts. Partly, this is simply because we spend a lot of time with ourselves, and so know more about our behaviour than anyone else. Another reason, however, is that our thoughts can be associated with our experiencing certain sensations, as well as engaging in certain behaviours (Ryle 1949; see also Schwitzgebel 2002). Consider Mark's belief that the No. 73 goes to Oxford Street. This belief might mean that Mark engages in certain patterns of behaviour, like walking to the bus stop. It might also mean that he experiences certain feelings. For instance, if Mark wants to get to a job interview in Oxford Street, he might feel a sharp pang in his stomach as he sees the No. 73 pull off from the bus stop without him. Since Mark, and nobody else, experiences this sensation, there is an important asymmetry between his position and anyone else's. Of course, a bystander might come to learn that Mark had this feeling, because he tells them so, or because they see him wince as the bus pulls away. But we have already conceded that people's knowledge of their own sensations is more immediate, and more secure, than anyone else's.

Does this show that we have a special means for finding out about our own thoughts? Not necessarily. The key point to notice is that, even when we are interested in our own patterns of behaviour and sensations, rather than someone else's, we still need to make sense of them. Often, these patterns are complex and conflicted, and the exact meaning of our behaviour, or even our sensations, is unclear to us. As Ryle puts it, 'pains do not arrive already hall-marked "rheumatic", nor do throbs arrive already hall-marked "compassionate"' (1949, 102). We might be perfectly well aware that we have experienced a particular sensation, like a pang in our stomach. But its precise meaning, and its relationship to our thoughts, might still elude us. Is the feeling in Mark's stomach a pang of anxiety because he thinks he'll be late for his interview? Is it a pang of sadness because he realizes that, if he gets the job, he'll have to move house to be closer to work? Or is it just a pang of ingestion because he rushed his breakfast? The answer will depend on other facts about Mark, such as whether he thinks another No. 73 will be along soon, whether he likes his house, or whether he did, in fact, rush his breakfast. To answer such questions, Mark must use the same set of shared, public concepts—concepts like belief, desire, and intention—that he uses to make sense of other people.

What all this means, I think, is that there *is* an asymmetry between our knowledge of our own thoughts and our knowledge of other people's. But this asymmetry arises simply because, in the case of our own thoughts, we tend to have more evidence available to us—including evidence concerning our sensations. It does not arise because we have any special means for finding out about our own thoughts, still less one that it is immune to error. To put it bluntly, there is no such thing as introspection, at least as far as our thoughts are concerned. Of course, this does not mean we cannot be introspective, in the ordinary sense of the word. We can pause to reflect on the things that we have said and done, as well as the feelings we had while doing them. In doing so, we can come to know our own beliefs, desires, or intentions. Often, our conclusions will be better informed than those that anyone else might draw—although sometimes a member of our family or a close friend might do better. We can grant all this. The point is only that, when we are being introspective, we do not gaze through an inner 'eye' and read off the contents of our thoughts, or have any other sort of direct access to them. Instead, we simply turn our ordinary folk psychological skills—the same skills that we use to interpret other people—upon ourselves, albeit with a bit more evidence to go on.

This view of self-knowledge is not new. Although many philosophers have claimed that we do have a special means of accessing our own thoughts, there are also those who have denied this. In fact, the view I have presented largely follows Ryle's remarks, and is similar to a position defended more recently by Peter Carruthers (2009, 2011). (See Byrne 2012 on the parallels between Ryle and Carruthers's accounts, and Wilkinson 2020 for helpful discussion and criticism.) Both Ryle and Carruthers agree that self-knowledge relies on the same folk psychological methods of interpretation that we apply to other people. Disagreement arises once we ask about the nature of these methods and the states that they seek to reveal. According to the fictionalist, our folk psychological concepts are metaphorical and governed by games of pretence. We are so thoroughly immersed in these games that they come to shape our view of ourselves just as much as they shape our view of other people. This is a familiar feature of games, of course. When children are caught up in a game of make-believe, its rules colour their understanding of their own actions and experiences as well as its props. Indeed, in Walton's terms, the children are *themselves* props in their games: Anna not only thinks *of the doll* as a baby; she thinks *of herself* as rocking the baby to sleep. The same happens in our folk psychological game. We make sense of other people by talking as if they were guided by an inner world of representations and we make sense of ourselves by talking as if we had such a world inside us. Once we are playing this game, it is only natural that we start to talk about self-knowledge as if it is a matter of peeking into this inner world (cf. Carruthers 2009, 127). As we have seen, there *is* a genuine asymmetry between our knowledge of ourselves and others. The metaphor of an inner world, and an inner eye that surveys its contents, give us one way to express this aspect of the mind. If we are not careful, however, it is a metaphor that can easily lead us astray.

56 MIND AS METAPHOR

2.3.2 Talking to Ourselves

There is one aspect of our mental lives that might seem to complicate the picture of self-knowledge that I gave in Section 2.3.1, and to cause particular trouble for mental fictionalism. That is our habit of talking to ourselves. Often, we talk to other people, of course. Sometimes, we tell them things that we already know about ourselves. Sometimes, we talk to them to figure out what we think about things, or what we want to do. We also talk to ourselves, however. I did this quite a lot when I learned to drive, telling myself things like 'OK, now—mirror, signal, check your blind spot' or 'Come on, clutch down!'. We also talk to ourselves to figure out what we think about things. I'll admit I've done some of this while writing this book. If I'm struggling to work out what I think about some knotty philosophical problem, I might try to talk it through with someone. If I can't find a willing victim, however, I might resort to talking to myself instead: I find myself trying to formulate the view I'm grasping for out loud, or voicing possible objections that a critic might put to me. Now, although it might be slightly embarrassing for me to admit this fact, it doesn't seem to cause trouble for the fictionalist. After all, speech—whether said to others or to ourselves—is still a form of behaviour. But now remember that, sometimes, we talk to ourselves silently, without making a sound—that is, we engage in what is called *inner speech* (or what Ryle calls 'silent soliloquy'). Thankfully for my passengers, I learned to do this while driving: I got to the stage where I could silently remind myself to check my blind spot. No doubt my colleagues are grateful that I don't always need to express my philosophical arguments out loud either. Instead, I find myself rehearsing arguments in inner speech and 'hearing' silent objections to them (typically as I lie in bed at night, trying to get to sleep).

The phenomenon of inner speech seems to threaten the view of self-knowledge that I presented in Section 2.3.1. I claimed that we have no special, error-proof means for finding out about our thoughts. Instead, we make sense of our behaviour and sensations in broadly the same way that we make sense of other people's. It is true that only we experience our sensations. But we cannot simply read off our thoughts from our sensations: we must still make sense of what we are feeling if we want to know what we believe, want, or intend. Inner speech might seem like an exception to this overall picture, however. Inner speech is commonly taken to be a sensory phenomenon. When engaged in inner speech, we have an experience that resembles hearing someone speak, even though there is no one speaking. Only we can 'hear' our inner speech. In this respect, it is like a minor ache or pain: it remains private to us, unless we choose to tell others about it, or our behaviour gives us away. And yet, unlike aches or pains, inner speech seems to be especially well-placed to give us access to our thoughts. Our episodes of inner speech seem to be more closely tied to the content of our thoughts. Indeed, it is tempting to say that, when we 'listen' to our inner speech, we are 'listening' our thoughts. When we are talking to ourselves in inner speech, we seem to have an immediate, conscious awareness of our thoughts, in a way that others do not.

What are we to make of this? The first point to notice is that, strictly speaking, episodes of inner speech are not thoughts, in the sense in which we have been using the term. This might sound odd. Terms like 'thought' (and 'thinking') are not precise. In everyday life, we might be happy to describe our experiences of inner speech as our 'thoughts'. In this book, however, I am using 'thought' in a more specific sense, to refer to our propositional attitudes (e.g. beliefs, desires, or intentions) as well as related aspects of the mind (e.g. concepts, reasoning, or understanding). A less misleading (and contentious) way to put the point, then, would be to say that episodes of inner speech are not propositional attitudes: they are not our beliefs, desires, or intentions (or our hopes, fears, decisions, judgements, etc.). This means that 'listening' to our inner speech does not allow us to 'listen' to our thoughts, in the sense of giving us any immediate access to our propositional attitudes. Instead, inner speech is much like our other sensations or behaviour: it simply provides us with a further body of evidence that must be interpreted if it is to tell us what we believe, desire, or intend.

To see why this is the case, recall some examples that we have encountered already. Suppose that I keep declaring how important it is to look after the environment, while still driving everywhere in my car when I could easily walk, dropping litter all over the place, and so on. It makes little difference whether I make my declarations out loud (either to myself or others), or whether I make them silently in inner speech. You would still be entitled to conclude that I don't believe what I say. Or consider Mark once again. Suppose that, in the first few years of working in his old job, he found that he would sometimes sit at his desk, saying 'I want a new job' silently to himself. If, despite this, Mark still arrived at the office each day with a smile on his face, sat down eagerly to begin work, and never made any attempt to consult the job adverts, we would be unlikely to say that he really wanted a new job. In fact, even Mark would be unlikely to say this, when he stopped to consider things. Taking the matter in the round, he'd probably put his moments of discontented musing down to a bad night's sleep, or the fact that he needed a holiday. He'd be unlikely to say that he'd intended to get a new job either.

It seems therefore that episodes of inner speech are only part of the evidence we use for interpreting someone's beliefs, desires, or intentions—even if that person is ourselves. Some thoughts might seem different, however. Beliefs, desires, and intentions are standing states: they are relatively long lived, and we are happy to say that we can be in these states even when we are not attending to them. For this reason, we might expect them to be somewhat removed from individual episodes of inner speech. In contrast, some thoughts seem more like events or episodes, which can be said to take place at a particular time. Consider decisions and judgements. Suppose that, in the first few years of working in his old job, Mark didn't intend to leave it, but now he does. At some point, it seems, Mark has taken

58   MIND AS METAPHOR

the decision to leave his job. We might even be able to specify the exact date and time when the decision took place—say, at midnight last New Year's Eve. If, while the chimes of Big Ben rang out, Mark said silently to himself 'I'm going to get a new job', wasn't he immediately aware of his decision? Similarly, if Mark now believes his job isn't paying him enough, when previously he was satisfied with his wages, there might have been a moment when he came to this judgement—say, a few minutes before midnight. If, at that moment, Mark 'heard' himself say 'they're not paying me enough' in inner speech, wasn't he immediately aware that he was making this judgement?

Despite appearances to the contrary, I think that our access to more fleeting thoughts, like decisions or judgements, is much the same as our access to standing states, like beliefs, desires, or intentions. Why is this? Fictionalism claims that the concepts of folk psychology are metaphorical, drawn from the world of public representations. In Chapter 1, we saw how this approach might be applied to thoughts like decisions and judgements, which are said to take place at a particular time. If a committee makes a decision, this occurs in public—or at least where its members can hear it. The committee deliberates over possible courses of action, declares its decision, and uses it to shape its future actions. Likewise, when an individual makes a decision, we talk as if a similar process goes on inside their head—a hidden, inner deliberation occurs and declares its conclusion, which shapes their future behaviour. In the case of judgements, our metaphor is drawn from the courtroom. A judge carefully weighs up the evidence before pronouncing her verdict, which is entered into the record. Similarly, when an individual makes a judgement, their inner courtroom weighs the evidence before pronouncing its verdict, which is entered into memory for future use.

Once we reflect on our concepts of decision or judgement, we can see that inner speech is no surer guide to these aspects of our mental lives than it is to our beliefs, desires, or intentions (see also Carruthers 2009, 133). The reason is that these concepts do not merely invoke the idea of inner representations; they invoke inner representations that play a particular role. In the case of decisions, they guide our future plans; in the case of judgements, they alter our beliefs. According to the fictionalist, of course, this is all just a convenient fiction. Someone who makes a decision merely behaves *as if* their inner committee had recorded a certain verdict, which they will now act upon. Likewise, someone who makes a judgement merely behaves *as if* their inner courtroom had pronounced on the matter and updated their records accordingly. For present purposes, however, the key point is that someone can easily rehearse a sentence in inner speech without having made the corresponding decision or judgement at all. If Mark 'hears' the sentence 'I'm going to get a new job' floating through his head as Big Ben chimes midnight, but then simply carries on behaving and feeling entirely as before (e.g. eagerly stating work each day, making no effort to search the job adverts, etc.), we would not say he had made the decision to get a new job. His intentions would remain as before.

Similarly, if he says silently to himself 'they're not paying me enough', but then continues to act and feel as he did previously (e.g. not asking for a pay rise, telling even close friends that he's happy with his wages, etc.), we would not say that he'd made this judgement. His beliefs would remain as before. When all is said and done, it was probably just the champagne talking.[1]

2.4 Conclusion

If Cartesianism takes a metaphor too seriously, anti-Cartesianism doesn't take it seriously enough. For this reason, well-known (and much criticized) anti-Cartesian positions like behaviourism and instrumentalism, as well as more sophisticated views like Ryle's and Dennett's, each tend to run into trouble. My aim in this chapter has been to show that, by acknowledging the metaphorical character of our talk about the mind, fictionalism can avoid some of these difficulties. It can also explain why folk psychology seems to treat mental states as causes and make sense of the way in which we find out about our own mental states. In the next chapter, we will turn to consider another key feature of our ordinary conception of mental states—namely, that they have content.

---

[1] My remarks on self-knowledge in Section 2.3 might go some way towards allaying a worry that Crane and Farkas (2022) raise for my earlier formulation of mental fictionalism in Toon (2016). As Crane and Farkas see it, even if fictionalism might be plausible for standing states, it is hard to see how we could doubt the existence of those thoughts that appear in our stream of consciousness. In essence, my suggestion is that there are no conscious thoughts, in the sense I have been using the term. Even when our stream of consciousness contains episodes of inner speech, the relationship between these episodes and our thoughts is a matter of interpretation. (For a similar view, see Carruthers 2009, 2011.)

3
Minds and Representations

In Chapters 1 and 2 I introduced mental fictionalism and showed how it differs from existing approaches to the mind. In this chapter, I will focus on one important feature of the mind, namely *intentionality*, or the mind's capacity to represent the world. The representational theory of mind tries to explain intentionality in terms of mental representations. On this view, we can think about the world because we have representations inside our heads that express the content of our thoughts. We have already seen that mental fictionalism denies that we have these inner representations. How then can fictionalism explain our capacity to think about the world? I will begin by briefly considering the ideas of intentionality (Section 3.1) and mental representation (Section 3.2). I will then develop an alternative, fictionalist approach to intentionality (Section 3.3). After doing so, I will be in a position to address two important objections to mental fictionalism: that it suffers from the problem of cognitive collapse (Section 3.4) and that the explanatory power of the idea of mental representations shows that these representations exist (Section 3.5).

3.1 Intentionality

We noted in Chapter 1 that our ordinary talk about the mind seems to take thoughts to be *about* things in the world. Our thoughts *represent* the world and have *contents*. My belief that there is tea in the tea caddy is *about* tea and the tea caddy. It *represents* the tea caddy: it represents it *as* having tea in it. The *content* of my belief is that there is tea in the tea caddy. In the philosophical jargon, thoughts have *intentionality* (or *aboutness*). There are other intentional phenomena too besides thoughts. Chief amongst them are spoken and written language. If I tell a colleague 'there is tea in the tea caddy' (or write her a note telling her this) then my words are also intentional. They too are about the tea caddy and represent it as having tea in it. In fact, they have the same content as my belief. Likewise, paintings, photographs, films, and plays all seem to possess intentionality—to say nothing of road signs, Roman numerals, semaphore, Morse code, and the rest. Each of these phenomena are capable of representing the world in one sense or another.

Among all these different intentional phenomena, some forms of intentionality seem to depend on others. Suppose that you and I have to go to a boring meeting

Mind as Metaphor: A Defence of Mental Fictionalism. Adam Toon, Oxford University Press. © Adam Toon 2023.
DOI: 10.1093/oso/9780198879626.003.0004

together. We agree on a signal beforehand: if I scratch my nose, you'll plead a prior engagement and give us an excuse to finish early. In this context, it seems that my act of scratching my nose has intentionality: it *means* something and has *content* (roughly, it means the same as the words 'plead a prior engagement!'). And yet its content would seem to stem entirely from our agreement beforehand. In this sort of way, some intentional phenomena (e.g. my signal when I scratch my nose) would seem to have a secondary status: their content is *derived* from that of other intentional phenomena (e.g. my words when I say 'if I scratch my nose, plead a prior engagement!'). We can therefore distinguish between *original* and *derived* intentionality. Phenomena with derived intentionality inherit their intentionality from other intentional phenomena, while those with original intentionality do not. Original intentionality must, it seems, be explained in some other way.

With all this in mind, we can ask a range of questions about intentionality (Haugeland 1990). Among all the intentional phenomena we encounter in the world, which of these have original intentionality and which are merely derived? How are we to explain original intentionality? And how is this original intentionality passed on to other intentional phenomena? My main aim in this chapter is to show how mental fictionalism approaches these questions. First, however, let us consider the approach taken by the representational theory of mind.

3.2 Mental Representations

According to the representational theory of mind, it is thoughts that possess original intentionality. All other intentional phenomena, including language, possess merely derived intentionality. To support this claim, representationalists often point out that a set of marks just sitting there on the page are not meaningful on their own. In itself, the word 'TEA' is just a smear of graphite or ink on some dried wood pulp. It is only because there are people around to use these marks—to read and write them, for example—that they come to have meaning. According to representationalists, the lesson that we should take from this observation is that the intentionality of language is derived from the intentionality of thought. Ultimately, it is because those who read and write the marks 'TEA' can have thoughts about tea that these marks mean what they do. Although the main idea behind this approach is clear enough, spelling it out in detail proves more challenging: exactly what thoughts (beliefs, intentions, and so on) must a speaker and hearer (or a writer and reader) have in order for a sentence to possess a particular meaning? Proponents of this approach must develop an account of this kind—often called a *Gricean* theory of meaning (see Grice 1989)—in order to explain how exactly it is that the intentionality of language is derived from thought.

On this approach, then, *all* intentionality—including that of language, pictures, diagrams, and any other sort of public representation—flows ultimately from the

62 MIND AS METAPHOR

intentionality of thought. It is our thoughts, and only our thoughts, that possess original intentionality. How can we explain the intentionality of thought itself? As we saw in Chapter 1, according to the representational theory of mind, being in a mental state is matter of having a representation inside your head with a certain content that plays the appropriate causal role. To believe that there is tea in the tea caddy is to have a mental representation with the content *there is tea in the tea caddy*. This mental representation is caused by certain sorts of experiences (e.g. seeing tea in the tea caddy) and guides your action in characteristic ways (e.g. reaching for the caddy to make a cup of tea). The story is similar if someone wants or hopes or (oddly enough) fears that there is tea in the tea caddy. Each of these thoughts gains its content from the relevant mental representation, each playing its own distinctive causal role in our mental machinery.

The upshot is that, according to the representational theory of mind, all intentional phenomena stem from the intentionality of thought and the intentionality of thought stems, in turn, from the content of mental representations. On this view, we can talk about the world (or draw it, map it, or paint it) because we can think about the world and we can think about the world because we have representations inside our heads that have content. At this point, of course, the natural question to ask is: how do these inner representations gain *their* content? How can part of our brain (or, if you prefer, part of our mind-stuff) *represent* tea or tea caddies? How can it be *about* these things, be true or false of them, accurate or inaccurate? In short, how can we explain mental representation? Broadly speaking, there would seem to be two main options open to us. First, we might accept that mental representation is a basic feature of the world, like charge or spin. Second, we might try to show how mental representation can be reduced to some other, non-intentional phenomena.

Most philosophers have found the first option unsatisfying, taking the view that, as Fodor puts it, 'if aboutness is real, it must be really something else' (1987, 97). They have therefore followed the second route, and sought to explain mental representation in non-intentional terms such as causal relations (Dretske 1981; Fodor 1987) or biological function (Millikan 1984; Papineau, 1987). One of the main challenges for such approaches is to explain the possibility of misrepresentation of various kinds. For example, a simple causal theory might say that a brain state represents some object in the world if and only if there is a reliable causal connection between the two, rather like the connection between smoke and fire. But this seems to rule out the possibility of misrepresentation: there may be no smoke without fire, but we can certainly *think* that there is a fire (or imagine or hope or fear that there is...) when in fact there isn't one. It is also difficult to see how this view could be extended to more complex concepts (e.g. democracy or charge) or explain our ability to think about things that lie outside the causal order (e.g. the number three or Count Dracula).

Despite initial progress in the 1980s and '90s, this project—the project of *naturalizing* mental representation—is commonly taken to have stalled somewhat (e.g. Godfrey-Smith 2005; Lycan 2022). Fictionalism points towards a very different approach to intentionality. Let us now turn to consider this approach.

3.3 Fictionalism and Intentionality

Recall that, according to the fictionalist, possessing a particular mental state is not a matter of having a representation with the right content inside your head. Instead, possessing a particular mental state is a matter of exhibiting a certain pattern of behaviour. The relevant pattern of behaviour is picked out using metaphors that are drawn from the world of public symbols, especially spoken and written language. To believe that the No. 73 bus goes to Oxford Street is, roughly speaking, to behave *as if* you had this written down in an inner notebook—although, of course, there is no such notebook. What does this mean for our approach to intentionality?

3.3.1 Thoughts

The first point to notice is that, for the fictionalist, our concepts concerning intentionality apply, first and foremost, to *public* representations, especially language. The core intentional notions are semantic categories such as meaning, truth, and reference. According to fictionalism, when we talk about mental states, we transfer these semantic categories metaphorically to the mind as an inner realm. Noting that people can say and write down sentences that have meaning— are about the world, are true or false, and so on—we begin to talk as if people also had such things inside their heads guiding their behaviour. An important consequence is that fictionalism presupposes that we can grasp these semantic notions without invoking mental states such as thoughts, beliefs, and intentions. This commitment is not unique to fictionalism, however. In fact, it lies at the heart of Sellars' myth about the origin of our talk about mental states. Our Rylean ancestors can engage in overt verbal behaviour before they have any notion of thoughts as inner episodes. It is only afterwards that Jones comes along and uses overt verbal behaviour as the model for his theory of our inner states.

Sellars' myth serves to highlight a further important point. For it shows that to say that our concepts concerning mental intentionality are based on those concerning public intentionality is not yet to say that the intentionality of the mental derives from that of public representations. To see this, notice that a representationalist might follow Sellars in thinking that public language serves as the model for our theory of inner states. And yet the representationalist will argue that these inner states exist. Using the model, we *discover* that people have these inner states and that it is these inner states that explain why people behave as they do— including their ability to use public language. In this way, the representationalist

64   MIND AS METAPHOR

will argue that, even if our concepts concerning public intentionality come first, it is still our inner representations that provide the basis for intentionality, including the intentionality of public representations. In a similar manner, someone might first possess a range of concepts relating to everyday objects like billiard balls (e.g. hardness, position, velocity), while knowing nothing whatsoever about atoms or molecules. In due course, however, they might use billiard balls as a model to construct a theory of atoms and molecules and conclude that it is these hidden particles that ultimately explain the behaviour of ordinary objects, including billiard balls.

Fictionalism takes a different route, however. For the fictionalist, public representations do not serve as a model for a new theory of the mind as an inner realm. Instead, they provide a fruitful source of metaphors for making sense of behaviour. Our semantic categories are only metaphorically, not literally, applied to the mind as an inner world. To say that someone has a certain mental state is not to say that they possess a particular inner representation; it is to say that they behave *as if* they had such an inner representation. For the fictionalist, then, the intentionality of mental states cannot be grounded in the intentionality of mental representations, since she argues that these inner representations do not exist; they are merely useful fictions. Instead, the intentionality of mental states is ultimately grounded in facts about a person's behaviour: it is because someone behaves in the way that they do that they can properly be said to possess certain mental states.

Of course, the fictionalist *also* claims that we make sense of people's behaviour using metaphors drawn from the world of public representations, such as notebooks, maps, and shopping lists. Notice, however, that this does not affect the fundamental point at issue—namely, that the intentionality of mental states is grounded in behaviour. The fact that we choose to describe someone's behaviour in a certain way does not, in itself, show that their behaviour is dependent upon our descriptions. For the fictionalist, a person possesses a given mental state in virtue of exhibiting the relevant pattern of behaviour, not because we pick out that pattern of behaviour using particular metaphors. As long as their behaviour remained the same, they would possess the same mental states—even if there was no one else around to see it. In virtue of their behaviour, Sellars' Rylean ancestors had *minds* even before Jones introduced his remarkable linguistic innovation—though this innovation might also have changed their behaviour in important respects too. (We will return to this point in Chapter 4.)

For the fictionalist, then, the intentionality of mental states is ultimately grounded not in mental representations, but in a person's behaviour. There is an obvious and important complication to this basic picture, however. For the behaviour associated with a particular mental state will often include the use of public representations, especially language. For example, one of the behaviours associated with believing that the No. 73 bus goes to Oxford Street is that, if someone asks you where the No. 73 goes, you say 'Oxford Street'. In some situations, you might write this down instead, or point to Oxford Street on a map, or indicate the relevant line in a timetable...and so on. This is not to say that mental states can be *reduced* to linguistic behaviour, of course. To believe *p* is not simply to be disposed to utter *p* (or write it down). As we saw in Chapter 2, fictionalism avoids the behaviourist's attempt to reduce talk about the mind to talk about behaviour, whether linguistic or otherwise. The point is simply that exhibiting the right pattern of behaviour to count as possessing a given mental state will often involve using language or other public representations.

Notice that I say that exhibiting the right pattern of behaviour will *often* involve the use of language, not that it will *always* do so. Consider animals and pre-linguistic infants. We often attribute mental states in such cases. We say that a dog knows where it lives or that a baby wants milk, even if neither can tell us as much. Fictionalism has no problem explaining such attributions. Consider memory and standing beliefs. The fictionalist claims that our attributions of standing beliefs are guided by the metaphor of memory as a notebook. This metaphor is most apt when applied to creatures that can use language. After all, people who use actual notebooks can typically read their contents if you ask them. But the metaphor can still be useful in other cases. In some respects, it is useful to treat a dog as if it had an inner notebook saying where it lives: it will help you to predict where the dog might end up if it gets lost. In other respects, of course, the metaphor is less apt: unlike someone who had such claims written down in a notebook, the dog cannot tell you its address or point it out on a map. The fictionalist need not insist on a sharp divide here, beyond which our metaphors cease to apply (see also Dennett 1996, 2013).

So the fictionalist need not claim that the ability to use language is necessary for the possession of mental states. For some mental states, like knowing where you live or wanting some milk, it is possible to exhibit some of the required pattern of behaviour without the use of language. For other mental states, however, the ability to use language is necessary. It is hard to know what a dog could do to convince us that it believes that current levels of inequality are a threat to liberal democracy, for example, or how a baby could show us that it wants the government to adopt a Keynesian economic policy. Let us mark this distinction by talking about *language-dependent* and *non-language-dependent* mental states. There will be considerable debate about which side of this divide certain mental states fall, of course. Can animals without language have intentions? Can they have reasons for their actions? Fortunately, we need not enter into these debates here. The important point is that fictionalism does not prejudge these issues by ruling out the possibility of minds without language.

3.3.2 Language

What about language and other public representations? How can we explain their content? It is important to acknowledge at the outset that fictionalism does not

66 MIND AS METAPHOR

provide a theory of public intentionality. It is not a theory of meaning. Fictionalism also places constraints on the theory of meaning that we might adopt. In particular, it will rule out any attempt to explain meaning in terms of the content of an accompanying mental state. For example, consider a simple Gricean theory that says that, in uttering U, a speaker S means p if and only if S utters U intending to produce in a hearer H the belief that p. Such a theory aims to reduce linguistic meaning to the contents of the mental states of the speaker and hearer. For those who adopt the representational theory of mind, this move is perfectly legitimate, since they will then look to explain the content of the speaker and hearer's mental states in terms of their inner representations. For the fictionalist, however, this move is not available and the theory risks circularity. According to the fictionalist, the speaker and hearer's mental states are grounded not in any inner representations, but in their overall pattern of behaviour. If this behaviour includes linguistic behaviour—if, that is, the relevant mental states are what I have called language-dependent—then our theory of meaning will be circular: we will have explained an utterance's meaning in terms of the speaker and hearer's mental states and then explained these mental states in terms of the speaker and hearer's utterances.

It would appear, then, that the fictionalist cannot explain the content of our sentences simply by appealing to the content of our mental states. At first glance, this might seem like an insurmountable problem. In Section 3.2, we noted that representationalists often defend their approach to intentionality by pointing out that a set of marks just sitting there on the page (or a ripple of disturbances in the air) seem to be entirely inert. From this, they conclude that we can only explain how such marks gain their meaning by appealing to the mental states of people who use them. If this is true, and if fictionalism cannot appeal to speakers' and hearers' mental states in this way, then it cannot explain meaning. Worse still, if fictionalism cannot explain meaning, its entire approach to intentionality is in trouble, because the fictionalist's analysis of mental states assumes that people can engage in meaningful linguistic behaviour. If there is no way to explain the intentionality of language without first assuming the intentionality of thought, then the whole approach threatens to collapse.

The situation is not quite as desperate as it might seem, however. There are other ways to explain the intentionality of language, besides appealing to the prior content of our mental states. The fictionalist can agree that, in themselves, marks on paper or sounds in the air are entirely inert or 'dead' (Wittgenstein 1953). Without people around, such marks would indeed be meaningless. It is only because they are taken up and used in certain practices that they come to possess meaning. For the fictionalist, the crucial point is that, when it comes to giving an account of these practices—that is, when it comes to explaining exactly how it is that the use of marks bestows meaning—our explanation must not rely upon the prior content of our mental states. Or, to be more precise, it must not rely upon the content of our language-dependent mental states—those mental states that already rely upon the use of language. Otherwise, our account will be circular. This still leaves a range of alternative explanations open to us, however. In particular, it leaves open the possibility that we might explain how meaning arises from norm-governed social practices.

What might such an explanation look like? Consider the following outline of one possible account (adapted from Haugeland 1990). Recall Sellars' mythical society. His Rylean ancestors were already able to use language, although this language was impoverished by its lack of psychological terms. According to the fictionalist, these Rylean ancestors already have (both language-dependent and non-language-dependent) mental states in virtue of patterns in their behaviour, even if Jones has yet to give them the metaphors to pick out these patterns. However, let us now consider a set of earlier 'pre-Rylean' ancestors, who lack the ability to use language. Since they cannot use language, the fictionalist must also conclude that the pre-Ryleans lack any language-dependent mental states. She cannot claim, as the representationalist might, that these thoughts are already lodged somewhere inside their heads, just waiting to be said out loud. The fictionalist can allow, however, that the pre-Ryleans have non-language-dependent mental states—the more basic beliefs, desires, and intentions that we might be willing to attribute to animals or pre-linguistic infants. The question now becomes: how might language and meaning arise in such a community?

Suppose that the pre-Ryleans are what Haugeland (1990) calls *conformists*: they tend to act alike, and to encourage others to act alike, rewarding them if they fall into line or punishing them if they don't. Within such a community, *norms* will arise: socially sanctioned forms of behaviour (*customs* or *practices*). Some of these practices will involve tools. Since practices are norm-governed, these tools have prescribed *roles*—ways in which they are *supposed* to be used. It is the role of a screwdriver, for example, to turn screws. This is what a screwdriver is *for*. The basic idea is that language itself can be understood a tool within these social practices. Some of the Ryleans' practices involve making certain sounds or marks while they do certain things. Like the screwdriver, these sounds or marks have prescribed roles within these practices: their use can be correct or incorrect, right or wrong, appropriate or inappropriate, and so on. Again, like the screwdriver, the sounds or marks are *for* something. For example, some of them might be for labelling or naming things. There might be many different language-using practices, however, and many different uses of language within them. To a first approximation, meaning *is* this norm-governed use of sounds and marks.

Of course, this is the briefest possible sketch of such an approach to language. There is much that must be done—and, indeed, has already been done—to develop and defend this sort of approach (e.g. Wittgenstein 1953; Brandom 1994; Hutto and Myin 2017). For our purposes, the important point is that fictionalism need not make meaning entirely mysterious. Instead, it presents us

68 MIND AS METAPHOR

with a challenge: if representationalism asks us to naturalize mental representation, fictionalism asks us to naturalize meaning. As we have noted already, fictionalism itself is not an answer to this challenge. Fictionalism is not a story about the pre-Ryleans and how they come to acquire language. Instead, it is a story about the Ryleans: it is a story about how, once a community has acquired language, it might then acquire the notion of mental states. According to the fictionalist, this happens when public intentionality is projected back on to a metaphorical inner realm. A community that already uses language as an *external* tool in its social practices—to name things, to make assertions, to ask questions—begins to talk as if it they had such things inside their heads.

3.3.3 Taking Stock

The upshot is an approach to intentionality that differs radically from that taken by the representational theory of mind. According to the representationalist, all intentionality stems ultimately from mental representations. For the fictionalist, there is no single source for intentionality. Instead, there are two: behaviour and social norms. The intentionality of the mental is grounded in behaviour. In its basic form, it can be possessed by creatures without language. In contrast, the intentionality of public representations, especially language itself, is grounded in social norms. These two forms of intentionality are fundamentally different, but also intricately related. With public representations come more complex forms of behaviour and, therefore, more complex mental states. Also with public representations come the metaphors we use to pick out the patterns of behaviour that ground our mental states. As Haugeland (1990) points out, any approach that admits the existence of two (or more) fundamentally different kinds of intentionality invites the question of what they have in common. Why call them both *intentionality*? For the fictionalist, the answer lies in the metaphors that figure in our ordinary talk about the mind. At the heart of folk psychology is not merely the notion of behaviour, but behaviour *as if* it were governed by inner representations. It is this metaphor that unites these two forms of intentionality, even if they are, in the end, fundamentally different phenomena.

3.4 Cognitive Collapse

We are now in a position to consider one of the most serious challenges facing mental fictionalism. That is the worry that it is incoherent. The problem is often put by saying that fictionalism suffers from 'cognitive collapse' (for discussion, see Joyce 2013; Parent 2013; Wallace 2016, 2022; Bourne and Caddick Bourne 2020;

Hutto 2022a). Let us first examine the nature of the problem (Section 3.4.1) before seeing how the fictionalist can overcome it (Section 3.4.2).

3.4.1 The Problem

The worry about cognitive collapse can be put as follows. Mental fictionalism argues that the inner states posited by folk psychology, like beliefs and desires, do not exist. According to the fictionalist, talk about such states is merely a useful fiction. And yet treating something as a fiction seems to require certain kinds of mental states: we are asked to *imagine* that what the fiction says is true or *make-believe* that such and such is the case. As a result, the critic alleges, mental fictionalism is incoherent: it denies the existence of mental states while assuming the existence of at least some of these states, such as imagination or make-believe. And if the fictionalist grants that *these* states exist, why deny the existence of other mental states? What justifies the unequal treatment? The worry can also be put in terms of representation. The fictionalist claims that inner representations do not exist and that our talk about them is a useful fiction. And yet a fiction is itself a representation: our folk psychological fiction, for example, *represents* people *as* having inner representations. To avoid collapsing like a house of cards, it seems that fictionalism must therefore allow that at least one representation exists and has content. Once again, if fictionalism allows this much, why deny the existence of mental representations in particular? Why the unequal treatment?

It might be tempting to dismiss this problem out of hand. As we noted in Chapter 1, it is misleading to say simply that fictionalism denies the existence of mental states. The fictionalist denies that we possess inner states that bear the contents of our thoughts and drive our behaviour. And yet she allows that there are real patterns in our behaviour that render our attributions of mental states true or false. In this sense, mental states are perfectly real. If that's right, perhaps there is no difficulty in the fictionalist appealing to imagination or make-believe after all? Sadly, things are not that easy. This response is correct as far as it goes, but it would leave fictionalism incomplete as an approach to the mind: it would mean that it could not be applied to certain mental states—namely, imagination or make-believe. The trouble is that fictionalism does not merely say that our attributions of mental states are grounded in behaviour; it offers an analysis of *how* this takes place. According to the fictionalist, we talk about behaviour via the fiction of inner representations. If our analysis of this process involves imagination or make-believe, then it cannot make sense of our attributions of *these* mental states themselves, or else it will be circular. The upshot is that, even if the fictionalist is entitled to assume the existence of mental states such as imagination or make-believe without fear of incoherence, she could not explain our talk about

70   MIND AS METAPHOR

them. The fictionalist analysis would have to be abandoned for such states. And if we were to abandon fictionalism for imagination or make-believe, the critic might insist, why not abandon it across the board? Once again, why treat these states differently?

The worry about cognitive collapse mirrors a well-known objection to eliminativism (e.g. Baker 1987). Like the fictionalist, the eliminativist argues that the inner states posited by folk psychology do not exist. And yet, the critic objects, asserting something involves believing it. So eliminativism is incoherent: the very act of asserting the position shows it to be false. Eliminativists are able to offer a compelling response to this objection, however. Paul Churchland (1981) argues that the charge of self-refutation begs the question against eliminativism: it assumes that we must explain what happens when someone makes an assertion (puts forward an argument, defends a position, etc.) in folk psychological terms. And yet this is exactly what the eliminativist denies. Eventually, according to eliminativism, we will come to possess a proper neuroscientific theory of activities such as asserting a position, putting forward an argument, and so on, and this theory will find no place for the categories of folk psychology.

The challenge facing mental fictionalism is more troubling than this objection against eliminativism, however. The incoherence facing eliminativism is alleged to arise not so much from the *content* of the eliminativist position as from the *act* of asserting it (Joyce 2013). Eliminativism is the claim that mental states do not exist. In itself, this position does not assume the existence of mental states; indeed, its sole claim is that they do not exist. Instead, the trouble is supposed to arise because the possibility of asserting any claim whatsoever—whether about philosophy or the weekend's football results—is said to require the existence of mental states. The eliminativist can justly reply that this assumption begs the question. The challenge facing fictionalism is more worrying, however, since the content of the fictionalist position *does* seem to assume the existence of mental states. The fictionalist does not simply claim that inner psychological states do not exist; she also tells us to regard talk about such states as a useful fiction. It is this further claim that generates the worry of inconsistency, for it seems to assume the existence of particular sorts of mental states, such as imagination or make-believe.

In this respect, the objection facing fictionalism is closer to that often levelled against Dennett's intentional stance (Adams and Aizawa 2001, 49; Bennett and Hacker 2003, 426). According to Dennett, when we attribute beliefs and desires to people, we do not claim that people have states bearing such content inside their heads. Instead, we adopt the intentional stance: we attribute those beliefs and desires that allow us to make sense of their behaviour. Here critics detect a problem. After all, adopting the intentional stance towards another person (or creature or artefact) would itself appear to be an intentional act: it is an interpretation that we use to make sense of their behaviour. The result is that Dennett is charged with much the same sort of incoherence as the fictionalist: the intentional stance would seem to assume the existence of exactly the phenomena that it denies exists. As Adams and Aizawa (2001, 49) put it, '[t]he content of Mike's attitude seems to depend on Ike's attitude, but whence comes the content of Ike's attitude?'

3.4.2 Pretending

To see how fictionalism can avoid cognitive collapse, we can begin by recalling a point we have already discussed in Section 3.3—namely, that fictionalism can acknowledge the existence of external, public representations with content.[1] As we saw, the fictionalist must deny that the content of such representations stems from the prior content of mental states. Instead, the most promising alternative looks to the role that these representations play within norm-governed social practices. Fictionalism is not itself a theory of this public form of intentionality, however; its scope is limited to the intentionality of the mental. It is here that fictionalism would seem to depart from Dennett's view. As I understand it, the intentional stance is intended to provide the whole story about intentionality. For instance, Dennett (2009, 345) rejects Robert Brandom's (1994) claim that only social creatures are capable of genuine belief. Instead, Dennett envisages a continuous spectrum of cases, from thermostats to Sherlock Holmes, with 'no theoretically motivated threshold distinguishing the "literal" from the "metaphorical", or merely "as if", cases' (2009, 343). As we saw in Section 3.3.1, the fictionalist can agree with Dennett that the intentionality of the *mental* is indeed a continuous spectrum: there is no clear line beyond which our metaphors cease to be apt. For the fictionalist, however, the intentionality of *public* representations stands apart: in this sense, fictionalism acknowledges that the emergence of social norms ushers in a new, and fundamentally distinct, form of intentionality. It is these 'literal' cases of intentionality that are projected back to yield the 'as if' intentionality of mental states.

How does this feature of mental fictionalism allow it to avoid cognitive collapse? The key point to notice is that, perhaps somewhat surprisingly, the fictionalist's analysis need not make essential reference to mental states like imagination or make-believe. Instead, its central notion is that of public, rule-governed acts of *pretence*. Recall the analysis I introduced in Chapter 1. I suggested that our ordinary attributions of mental states are metaphorical and that we should analyse these metaphors using Walton's account of prop-oriented make-believe. At the heart of this analysis is the idea that, in 'pretending to say one thing, one may

---

[1] Note that the response to the problem of cognitive collapse that I give in Section 3.4.2 differs from that mooted in Toon (2016), which I no longer think is successful. For critical discussion, see Bourne and Caddick Bourne (2020).

72 MIND AS METAPHOR

actually be saying, asserting, something else' (Walton 2000, 95). In the specific case of mental fictionalism, the core idea is that, in pretending to describe inner representations, we are actually saying something about behaviour. This takes place within the context of folk psychology, which is a particular kind of rule-governed social practice. To avoid the charge of incoherence, the fictionalist must show that we can make sense of this notion of pretence without reference to mental states.

This might seem like a tall order. At first glance, no doubt many would be tempted to agree with Lynne Rudder Baker when she writes that 'it is difficult to see how even to understand pretense without presupposing belief. To pretend that this cup contains water seems to entail believing that it does not' (1994, 199). On closer inspection, however, the link between pretence and (dis-)belief is less straightforward than it might seem. For example, surely children playing bears in the woods believe that bears growl and gnash their teeth, as well as pretending that they do (Walton 1990, 13; Ryle 1949, 244; for further discussion, see Langland-Hassan 2014.) Another natural thought is that, in order to count as pretence, an action must be accompanied by some private, inner act of imagination. As Ryle observes, however, this claim is also implausible. Suppose that we come across a child who is padding around the woods, growling, and gnashing his teeth (Ryle 1949, 243). To judge whether he is pretending to be a bear, we do not need to ask whether he is also undergoing some hidden, inner imaginative act of silent growling or gnashing of teeth. His public acts of growling and gnashing are enough.

How else might we understand pretence, if not through its connection to mental states such as belief or imagination? The alternative, I suggest, is to characterize pretence as a form of public representation. Given enough practice, a child's growl might sound indistinguishable from a bear's growl. And yet the two are very different sorts of act. The bear's growl is merely a growl but the child's growl is, at one and the same time, an act of representation (Ryle 1949, 245–6). The child's growl only makes sense in light of the bear's, whereas the reverse is not true. Similarly, a quotation might be indistinguishable from the passage it quotes. And yet the quotation remains a very different thing to the original: the original sentence might have been uttered in an impassioned political speech while the quotation appeared in a dry newspaper report. The difference between the two will not be revealed by close scrutiny of the sentence itself or the mental states of the politician or newspaper reporter. Instead, the difference lies in understanding public speaking and newspaper reporting as distinctive social practices with their own particular norms. In a similar vein, we may characterize acts of pretence not in terms of any connection to mental states like belief or imagination, but as a distinctive form of external, public representation that occurs within its own kind of social practice—namely, the practice of playing games.

To see how this might work, consider assertion. On the face of it, we might wonder how we could understand assertion except by referring to mental states.

MINDS AND REPRESENTATIONS   73

Isn't asserting something a matter of saying what you believe? Another way to understand assertion, however, is to see it as a distinctive sort of norm-governed social practice—what Sellars called 'the game of giving and asking for reasons' (Brandom 1997, 123). Here is a (much simplified) example to illustrate the basic idea. Suppose that John says, 'the baby is asleep'. There are certain circumstances in which it is appropriate for John to say this (e.g. if he's standing in front of the cot, watching the baby sleep) and others in which it is not (e.g. if he's standing in front of the cot, hearing the baby scream blue murder). Likewise, there are certain further actions that are appropriate given what John says (e.g. saying that someone is asleep, tiptoeing around the room) and some that are not (e.g. saying that everyone is awake, playing the trombone). Because these norms are in force, John has certain entitlements and commitments within the 'game' of assertion. He is entitled to engage in further actions (e.g. saying that someone is asleep), but also committed to defending what he says (e.g. if his wife says she can hear a baby crying, he might point out that the sound is coming from next door). Roughly speaking, the idea is that making an assertion is a matter of engaging in a social practice that is governed by these sorts of norms.

My suggestion is that we might understand pretence in a similar manner. Like asserting, pretending is a matter of engaging in a certain sort of norm-governed social practice—namely, games. The norms that govern these games are different from those that govern the 'game' of assertion, of course. Let us return to our example of Anna and David playing with dolls. Suppose that Anna says, 'the baby is asleep'. There are certain circumstances in which it is appropriate for Anna to say this (e.g. she's looking at a doll and can see that its 'eyes' are closed) and others in which it is not (e.g. she's bouncing the doll on her knee and its 'eyes' are open). Likewise, there are certain further actions that are appropriate given what Anna says (e.g. saying that someone is asleep, tiptoeing around the room) and some that are not (e.g. saying that everyone is awake, playing the toy trombone). Because these norms are in force, Anna has certain entitlements and commitments within her and David's game. She is entitled to engage in further actions (e.g. saying that someone is asleep), but might also be called upon to defend what she says (e.g. if David wonders whether the dolls 'eyes' are really closed, she might show him that they are). Again, the idea is that engaging in pretence is a matter of engaging in a certain sort of norm-governed social practice—in this case, a game in the ordinary sense of the word.

There is an immediate objection facing this approach to pretence, however (Langland-Hassan 2022). After all, there are lots of different sorts of games, but not all of them involve pretence. Think of football, snooker, or playing tag in the playground. How might we narrow our focus to exclude these cases? Can we say what is distinctive about games of pretence, or the norms found within them? So far, we might have been content to say that these games involve norms that prescribe what we are supposed to pretend. But we cannot say that here, of course.

74 MIND AS METAPHOR

Our aim now is to ask whether we can characterize pretence by pointing to its role in certain norm-governed activities. Clearly, if we try to pick out those activities by helping ourselves to the notion of pretence, we will not have got very far. Can we find another way to pick out these games? This will certainly not be easy. Wittgenstein (1953) famously pointed out the sheer variety of activities that we call games. This might make us wary of trying to capture something is common to all games of pretence (Hutto 2022b). Still, perhaps we might try to identify some distinctive characteristics shared by (at least some) central examples of games of pretence, which are not found in other games.

A good place to start is to recall a point that we noted earlier—namely, that acts of pretence involve a certain sort of representation. A bear's growl is merely a growl. But if a child growls while playing at bears, the noise they make is different: it depends on the bear's growl; it only makes sense in light of the bear's growl. The same is true of games with dolls. Suppose that John rocks the baby back and forth in his arms to get it to sleep. Playing with a doll, David might make the same motion, holding the lump of plastic in his arms and swaying gently back and forth. Just as the child's growl depends upon the bear's, so David's action depends on actions like John's; it only makes sense in light of the fact that this is how people (sometimes!) manage to get babies to sleep. More generally, the games that children play with dolls depend upon our ordinary practices for looking after babies: the things that children do and say in these games only make sense in light of those ordinary practices. Speaking loosely, we might say that the norms that govern children's games with dolls 'mirror' (albeit imperfectly) the norms that govern our practices for looking after babies. To a certain (limited) extent, when children play with dolls, they transfer the norms that guide our interactions with babies over to their interactions with a lump of plastic. In this way, games of pretence 'look beyond' themselves to other objects, events, and practices. Other sorts of games seem to lack this 'higher order' character. The rules of tag stand on their own. They do not depend upon other activities or practices (e.g. catching bears) for their significance. If they did, then we would be inclined to count them as games of pretence.

The basic idea, then, is that we might try to pick out games of pretence by their distinctive higher order character. A distinguishing feature of (at least some) central examples of games of pretence (e.g. playing with dolls) can be found in the relationship that they bear to other practices (e.g. looking after babies). When we play with a doll, the norms that guide our interaction with one thing (i.e. a lump of plastic) 'mirror', or are transferred from, the norms that shape our interactions with something else (i.e. a baby). To pretend that a lump of plastic is a baby—to treat a lump of plastic 'as if' it were a baby—is to take part in one of these sorts of games.

This is quite rough as it stands, of course. There is much that would need to be done to refine and develop this approach, and to show that it can explain different sorts of pretence. Still, it seems to give us the resources that we need to explain some important features of pretence, without appealing to mental states like belief or imagination. For example, it can help to explain the difference between acts of pretence and mere accidents or mistakes. A man who is left limping after a football match might happen to walk like a pirate with a wooden leg, but he is not pretending to be a pirate (Langland-Hassan 2014). Similarly, a drunk who mistakenly picks up a banana instead of the phone is not pretending that the banana is a phone (Langland-Hassan 2022). The reason that these are not cases of pretence, I suggest, is that neither the footballer nor the drunk are playing a game. Although they make the right bodily movements, neither is taking part in the right kind of norm-governed social activity that would be required for their actions to count as pretence. Similarly, if I am walking in the park, I might accidentally stray onto a football pitch while a penalty is being taken. The penalty taker strikes the ball cleanly, but rather than flying into the top corner, it hits me instead. My bodily movement might be much the same as the goalkeeper's (if I hadn't got in her way). Still, my actions would not count as saving a penalty. Why not? Because I am not taking part in the game. On the contrary, I've stopped the game.

What does this all mean for mental fictionalism? Does understanding pretence in this way allow us to avoid cognitive collapse? We have seen already (in Section 3.3) that the fictionalist must assume that our concepts concerning public intentionality are prior to those concerning mental states. This is part and parcel of Sellars' myth. Our Rylean ancestors can talk before Jones comes along; they already possess concepts of meaning and truth, as well as related ideas such as asserting, questioning, promising, and so on. To avoid cognitive collapse, the fictionalist must insist that the Ryleans could also engage in pretence. As well as making assertions or asking questions, they could also play games with dolls, put on plays, or tell stories. If this is an embellishment of the myth, it is a fairly minor one. In this section, I have suggested one way in which we might begin to construct an account of pretence by appealing to social norms. Of course, this suggestion would need to be developed further to provide a fully fledged theory of pretence. Notice, however, that the details of this theory will not depend upon mental fictionalism. Fictionalism itself does not provide a theory of pretence, much in the same way that it does not provide a theory of meaning or assertion. That is hardly surprising, of course. It shows only what we have seen already— namely that fictionalism, if taken as a general theory of intentionality, is incomplete, not that it is incoherent.

3.5 Why Not Representationalism?

Advocates of the representational theory of mind will be growing impatient. If our talk about the mind involves talk about inner representations, why not take it at

76 MIND AS METAPHOR

face value? And if such talk is so successful in explaining and predicting people's behaviour, why not conclude that such representations exist? In a sense, this line of argument can be levelled against all anti-Cartesian approaches to the mind, including behaviourism or instrumentalism. But it is especially troubling for the fictionalist. Unlike other anti-Cartesian views, fictionalism grants that talk about the mind involves talk about inner representations. The fictionalist simply argues that such talk should be understood as pretence. And yet, the representationalist might ask, if such talk is so useful, why not take it more seriously? In short, why not accept representationalism? Let us turn to consider this objection. In doing so, we will need to reflect more widely on the role of fictions in our thought.

3.5.1 Fictions and Hypotheses

In his classic work on fictions, *The Philosophy of 'As If'* (1924), Hans Vaihinger makes a useful distinction between *fictions* and *hypotheses*. For Vaihinger, *fictions* are false and taken to be false by those who use them. The real test for a fiction is not whether it is true, but whether it is useful. By contrast, *hypotheses* make claims about the world whose truth remains an open question, which we hope to settle by future inquiry. Consider the billiard ball model of gases. The model makes use of many fictions, in Vaihinger's sense. It assumes that the molecules of the gas collide perfectly elastically, for example, and that they exert no forces on each other between collisions. These assumptions are false and known to be so. And yet they are also enormously useful, since making these assumptions allows us to construct our model of the gas and its behaviour. The model also involves a number of hypotheses. For example, it claims that the gas is made up of molecules and that they exert pressure when they collide with the walls of the container. These are claims about the gas and its behaviour that might have turned out to be true or false—although, in this case, we believe that these hypotheses are true.

The distinction between fiction and hypothesis is primarily a matter of the attitude that we take towards a particular statement, rather than the statement itself. At one time, the Ptolemaic system might have been intended as a hypothesis about the workings of the universe: if so, it was a hypothesis that turned out to be false. And yet even today an astronomer might still employ the Ptolemaic system as a useful fiction to calculate the positions of the stars in the night sky. In this way, a statement might change its status from hypothesis to fiction. In principle, a statement might also move in the opposite direction, from fiction to hypothesis. The development of atomism in nineteenth-century chemistry might offer an example of this. As late as the 1860s, many chemists remained sceptical, or at least agnostic, regarding the existence of atoms in the sense we would recognize today. Although models of chemical structure were useful for explaining isomerism, chemists insisted that these models were not to be misunderstood as showing the actual positions of atoms in molecules. Towards the end of the century, however, the very same models were taken to show exactly that. (Of course, the details of these developments are somewhat more complicated. For a detailed historical account, see Ramberg 2003; see also the discussion in Toon 2012, ch. 4.)

The representationalist's objection can now be put as follows: why not treat talk about mental representations not as a fiction, but as a hypothesis? Moreover, why not treat it as a hypothesis that is likely to be true? In Chapter 1, I argued that the folk regard talk about inner representations as a useful fiction and that this interpretation fits well with a number of puzzling features of our ordinary talk about the mind. For example, it fits well with the fact that it makes little sense to count beliefs or ask where they are located. The representationalist might concede this point. Perhaps the folk do regard mental representations merely as useful fictions. Maybe *they* do not care much about people's inner workings. But what if we are more curious about what goes on inside people's heads? Why not take such talk seriously? Why not take it as providing us with a fruitful set of hypotheses about the workings of the mind (cf. Bloor 1970)? Since these hypotheses provide our best explanation for people's behaviour, we have good reason to think that they are true. Indeed, if they weren't true, then the success of these hypotheses in predicting people's behaviour would be a miracle.

We need to tread carefully here. Suppose that we do treat talk about mental representations as a hypothesis. Why think this hypothesis is likely to be true? Here the representationalist is making two well-known realist manoeuvres. The first is *inference to the best explanation* (Lipton 2004). According to inference to the best explanation, we are entitled to infer the hypothesis that would, if true, provide the best explanation for the evidence. The realist argues that our hypothesis involving mental representations provides the best explanation for why people (and perhaps also animals) behave in the way that they do. According to inference to the best explanation, we are therefore entitled to infer that this hypothesis is true. The realist's second manoeuvre is the *no miracles argument* (Putnam 1981). According to the no miracles argument, if a theory is successful, the best explanation for its success is that it is true (or at least approximately true). If it weren't true, the realist claims, then its success would be a miracle. So we are entitled to take successful theories to be true. The no miracles argument also involves an inference to the best explanation, this time at a higher level than ordinary scientific inference. The scientist argues that a theory is true because it best explains the evidence (e.g. people's behaviour). The realist then argues that the theory is true because truth is the best explanation for its success (e.g. its accurate predictions of people's behaviour).

Much has been said about both of these realist strategies. Not all will accept inference to the best explanation as a form of inference (Van Fraassen 1980) and even those who do accept it might question the cogency of the no miracles argument in particular (Lipton 2004). In the present context, the important

78 MIND AS METAPHOR

point is that both strategies begin to look rather less straightforward once we pay close attention to the role played by fictions in science (Fine 1993; see also Suarez 2009). Typically, when scientists want to explain or predict the behaviour of some part of the world, they must first construct a model (e.g. Cartwright 1983; Giere 1988; Morgan and Morrison 1999). All models involve fictions, in Vaihinger's sense: they deliberately simplify or idealize a complex, real world phenomenon in order to make it more amenable to theoretical treatment. This threatens to cause difficulties for both of the standard realist arguments that we have considered. For it now seems as if the best accounts that scientists can offer for explaining a phenomenon or predicting its behaviour are often false and known to be so. What does this mean for inference to the best explanation or the no miracles argument?

The issues here are complex and have yet to be fully explored (for a helpful overview, see Levy 2018). At least this much seems clear, however. The widespread use of fictions in science shows that, rather than making a blanket appeal to inference to the best explanation or the no miracles argument, realists must pay close attention to the particular models at issue. Suppose that we model a swinging weight as a simple harmonic oscillator. In doing so, we invoke a number of fictions: we treat the weight as if it were a point mass, treat the string as if it were massless, ignore air resistance, and so on. Now let us assume that, despite these simplifications and idealizations, our model nevertheless explains the motion of the weight and yields accurate predictions. What should the realist say about this? She might justifiably claim that our model's success entitles us to think that it has captured *some* underlying truth about the swinging weight and its behaviour. But its success does not show that the weight is a point mass after all, or that the string really is massless, since we know perfectly well that this is not the case. When we have good reasons to think that elements in a model are fictions, its success will not normally lead us to conclude that these elements are true after all.

This is not to say that fictions can never become hypotheses. As we have seen, nineteenth-century chemistry might offer an example of this. The point is simply that we often have good reasons to think that certain elements in a model are fictions and the model's success will not lead us to revise our opinion. How do we know that a particular element in our model is a fiction? Here it is helpful to draw on another distinction due to Vaihinger (1924), between what he calls (slight confusingly) *semi-fictions* and *real fictions*. The key difference is that, while *semi-fictions* are false and known to be so, *real fictions* are not merely false but incoherent. For instance, if we ignore the effects of friction, we invoke a semi-fiction: we know our assumption is false, but there is nothing incoherent in the idea of an object moving without encountering friction. Other fictions are more puzzling. Discussing atoms in nineteenth-century physics, Vaihinger (1924, 218–19) writes that:

> If ... we designate the atoms as centres without extension, we are merely creating substantial basis for the relationships of force, a basis that, upon more accurate

scrutiny, turns out to be a very strange construction indeed. For an entity without extension that is at the same time a substantial bearer of forces—this is simply a combination of words with which no definite meaning can be connected.

For Vaihinger, the notion of atoms as centres without extension is a real fiction: when we stop to think about it, we simply cannot make sense of the idea that all the mass of an object could be located at a point without extension. And yet this does not prevent the idea playing an important role in physical theory. As Vaihinger puts it, 'the concept in question is contradictory, but necessary' (1924, 72).

Whether a claim is a semi-fiction or real fiction will obviously have a bearing on our reasons for taking it to be a fiction, rather than a hypothesis. In the case of semi-fictions, we typically regard a claim as fiction because it runs counter to some other claim about the world that we hold to be more certain. For example, we take our assumption about air resistance to be a (semi-)fiction since we have independent reasons for thinking that objects moving in the atmosphere encounter air resistance. The success of our model will not lead us to revise this view—although it might reassure us that we were correct to think that we could safely ignore air resistance in this instance. In the case of real fictions, however, we can offer a rather different reason for thinking that we are dealing with a fiction. After all, if a claim is incoherent, then it cannot be true. Consider our assumption that the swinging weight is a point mass. Of course, there is a sense in which we can simply look and *see* that it is not a point mass. If Vaihinger is correct, however, then we have a stronger reason for thinking that we are dealing with a fiction: whatever else we know about the swinging weight, if our assumption is incoherent, it simply cannot be true.

3.5.2 Mental Representations as Real Fictions

Why does the fictionalist regard talk about inner representation as a fiction, rather than a hypothesis? We certainly cannot simply look and see that people do not have such things inside their heads. The reason that we should regard mental representations as fictions, I suggest, is that they are real fictions, in Vaihinger's sense: the notion of mental representation that we find in folk psychology is not merely false, but incoherent (cf. Appiah 2017). Our ordinary concept of representation concerns external, public forms of representation that represent through social conventions, like spoken or written language, maps, or diagrams. And yet it is clear that mental representations cannot be representations in this sense, since they are supposed to be locked away inside people's heads and are certainly not subject to any social conventions. For the fictionalist, mental representations are much like point masses or massless strings. Talking as if the mind were an inner world containing such representations is an enormously productive way of

80 MIND AS METAPHOR

making sense of people's behaviour. And yet, when we stop to think about it, we see that the idea that people could really such things inside their heads makes little sense (cf. Sprevak 2013).

In this context, we might recall Wittgenstein's famous remark that 'only of a living human being and what resembles (behaves like) a living human being can one say: it has sensations; it sees; is blind; hears; is deaf; is conscious or unconscious' (1953, 281). This remark has inspired Maxwell Bennett and Peter Hacker (2003) to argue that neuroscientists' talk about the brain containing beliefs, knowledge or, indeed, representations is simply incoherent. After all, a brain does not resemble a living human being. On this central point, the fictionalist can agree. Notice, however, that immediately after making this remark, Wittgenstein considers an objection: '"But in a fairy tale the pot too can see and hear!"' (1953, 282). If we tell stories like this, doesn't this show that we *can* make sense of attributing mental states to inanimate objects after all? The same point can be made about mental representations. Philosophers often dream up thought experiments about creatures with sentences or pictures inside their heads (e.g. Sprevak 2009). Indeed, the fictionalist must rely on this fact: we must be able to *pretend* that people possess such inner representations, even if they do not. If we can engage in this pretence, doesn't this show that talk about mental representations *is* coherent after all?

This would be too quick. In the first place, notice that a fairy tale will usually have us imagine that a pot can see or hear by making it *behave* like a human being in certain respects—it might be able to shout or run away if someone tries to fill it with hot water (Wittgenstein 1953, 282; see also McGinn 1997). If *this* is what we imagine, then the fairy tale will not show that it makes sense to attribute mental states to inanimate objects, since the pot is not inanimate. A similar lesson applies in the case of mental representations. When we are asked to imagine a creature with inner representations, I think, we often find ourselves imagining an inner agent (or homunculus) who *reads* these representations. If *this* is what we imagine, then such thought experiments do not show that it makes sense to talk about mental representations. This scenario might well be coherent, but it is not what the representationalist needs. After all, most will not want to countenance inner homunculi.

More importantly, though, it is simply false to say that, whenever we engage in pretence, our pretence must be coherent—if, by 'coherent', we mean that it must make sense if taken literally. In fact, we often have little idea what it would mean to take our pretence literally. Consider the fairy tale with the talking pot. Wittgenstein asks, 'Is it false or nonsensical to say that a pot talks? Have we a clear picture of the circumstances in which we should say of a pot that it talked?' (1953, 282). Arguably, we do not. Our use of language in fairy tales stands somewhat apart from that of ordinary life, and it can be difficult to see what it could mean to take it literally. Think of children's games. In the midst of playing a game, children will suddenly find they can perform amazing feats of magic, turn themselves into ghosts or monsters, disappear into thin air, travel back in time, or turn into a rocket and launch themselves into space. In many cases, it is hard to see what it could mean to take these ideas seriously. And yet we are still able to engage in these sorts of pretence perfectly well. Indeed, even small children can play games like these.

The same is true in the case of metaphors. Some metaphors can be understood literally. People can literally, as well as metaphorically, find themselves standing at a crossroads or carrying a heavy burden. Many metaphors do not make sense if taken literally, however (Walton 2000, p. 96). It is hard to know what it would mean to say that clouds are literally, rather than metaphorically, angry. Our ordinary concept of anger simply does not apply to objects like clouds. And yet we can still use this idea figuratively. Saying that clouds are metaphorically, rather than literally, angry is perfectly meaningful. It can even help us to pick out genuine facts about the state of the weather today. Many metaphors that we use to describe the mind are like this: they do not make sense if taken literally. Chief amongst these, I suggest, is talk about inner representations. Our ordinary concept of representation simply does not apply to representations inside the head. And yet we can still use this idea figuratively. Our metaphor is perfectly meaningful. As we have seen, it can even help us to pick out genuine facts about people's behaviour.

For the fictionalist, the important point is that, even if the notion of inner representations cannot be understood literally, we can still pretend that people have such things. The notion of representation has a literal use when applied to external forms of representation, like notebooks, maps, or to do lists. It also has a metaphorical use when applied to the mind as an inner world. Both of these uses are legitimate. The trouble arises only when we begin to confuse them.

Three caveats are in order at this point, however.

First, our discussion has focused on the notion of mental representation as it appears in ordinary talk about the mind. It is this notion that I have argued is a real fiction. The fictionalist need not deny that proponents of the representational theory of mind might one day develop a new, technical notion of representation that does not suffer from this incoherence. In a sense, of course, this is precisely what representationalists have tried to do: they have tried to show how inner representations might gain their content through some other means, such as causal relations or evolutionary history. Up till now, this project has not been successful, but we should not rule out the possibility that it might succeed one day. Even if this project is ultimately successful, however, we might still ask what this technical notion has in common with our ordinary concept of representation, or what the existence of such representations tells us about our ordinary concept of mind. Certainly, if the fictionalist's analysis of our ordinary talk about the mind is along the right lines, it is debatable whether these inner representations—even if they do exist—would count as beliefs or desires.

82 MIND AS METAPHOR

Second, even if the notion of mental representation cannot be taken literally, this does not mean that talk about them cannot do useful work, even in scientific contexts. Opponents of the idea of mental representation often assume that, if attributions of inner representations cannot be true, then they can play no useful role in cognitive science—at least, no useful explanatory role (e.g. Bennett and Hacker 2007, 140; cf. Dennett 2007). And yet, as we have seen, *all* scientific models make assumptions that we know are not true, while others invoke ideas that are hard to take literally, like point masses, massless strings, or infinite potential wells. If we insist that truth is necessary for explanation, we risk finding that science has delivered us far fewer explanations that we might have hoped. Of course, the fictionalist can agree that talk of inner representations must be handled with care: we must not take our metaphors *too* seriously. Still, we must take them seriously enough if we are not to overlook the vital role they play in much of our thinking. Surveying debates over the notion of the atom in nineteenth-century physics, Vaihinger writes,

> The defence was always anxious to show that the alleged contradictions were only apparent and that the concept therefore possessed objective validity and could be applied. Their opponents, on the other hand, demonstrated the contradictions and so refused to allow the concept any legitimate place in science; in other words, they poured out the baby with the bath, while the defence accepted it—un-washed.   (1924, 71–2)

Much the same could be said about contemporary debates over mental representation.

Third, notice that fictionalism—at least as I've presented the view—need not imply any wholesale opposition to realism. In particular, it need not impose a blanket ban on inference to the best explanation (cf. Sprevak 2013). I've suggested that there is a particular reason why the success of folk psychology should not lead us to infer the existence of mental representations: these inner representations are real fictions and so our folk stories about them cannot be true. This need not threaten inference to the best explanation, which requires that the best explanation is *good enough* to warrant our inference (Lipton 2004). At a minimum, this requires that it is coherent. This is not to say that fictions cannot explain but only to say that, if they are incoherent, they cannot even be candidates for inference to the best explanation. After all, inference to the best explanation tells us to infer what would, *if true*, provide the best explanation for the evidence—and real fictions cannot be true (see also Levy 2018). So fictionalism need not reject inference to the best explanation. It should, however, lead us to reject any defence of representationalism that appeals merely to the success of folk psychology. It is often said that, since folk psychology is successful, we can be confident that mental representations exist, even without any naturalistic account of mental representation in hand. This is too quick. Before we can appeal to the success of folk psychology to argue for the existence of mental representations, we must first show that we are not dealing with a real fiction. After all, our physical theories might enjoy any number of successes, but we do not take this to demonstrate the existence of point masses or massless strings.

For the fictionalist, then, the success of folk psychology does not entitle us to infer its truth, since the story it tells about the mind simply cannot be true. Fodor (1975) famously defended representationalism on the grounds that it is the only game in town. In a sense, the fictionalist can agree. The trouble is that, like many games, representationalism is hard to take seriously. And yet, even if we accept this point, we might still feel as if we are left with a mystery that has needs to be solved. After all, if the truth of folk psychology does not explain its success, then what does explain it? Why on earth should a story that isn't true—indeed, a story that I have said is incoherent—allow us to explain and predict people's behaviour? Does fictionalism make the success of folk psychology a miracle? We will return to this worry in Chapter 4.

3.6 Conclusion

Fictionalism is popular in many areas, from mathematics to morality. Mental fictionalism has few adherents, however. Much of the blame lies with the problem of cognitive collapse: if mental fictionalism assumes the existence of the very thing that it brands a fiction, it is hard to see how it could even get off the ground. In this chapter, I hope to have shown that the situation is not as dire as it seems. Properly understood, mental fictionalism suggests a new approach to intentionality that is coherent, if incomplete. In the beginning was the deed. After that, came the word—which brought with it new deeds and new ways to describe them. Our ordinary talk about the mind is a metaphorical mapping of words onto deeds.

4

Minds and Materials

What is the relationship between mental states and items of material culture, like notebooks, maps, or lists? The extended mind thesis (ExM) offers an influential and controversial answer to this question. According to ExM, our mental states can be partly constituted by items of material culture: together with our brains, things like notebooks, maps, or lists can form part of the physical stuff that realizes our beliefs and desires. Although ExM offers a radical view of the *location* of mental states, it fits comfortably with a traditional, representationalist account of the *nature* of those states. In this chapter, I will offer a fictionalist analysis of the relationship between mental states and material culture. After introducing ExM and its relationship to representationalism (Section 4.1), I will show that fictionalism yields a fresh approach to the idea of extended mental states (Section 4.2). I will then show how this approach avoids well-known objections to ExM (Section 4.3). At the end of the chapter, I will return to the worry that fictionalism makes the success of folk psychology a miracle (Section 4.4).

4.1 Material Culture

Let us begin by considering our interactions with material culture (Section 4.1.1), before turning to examine the extended thesis (Section 4.1.2), and its relationship to representationalism (Section 4.1.3).

4.1.1 Thoughts and Things

We have already encountered at least two important ways in which, according to a fictionalist approach, minds and material culture are intertwined. First, material culture is the source of many of our metaphors for talking about the mind. In Chapter 1, for instance, I suggested that we talk about memory as if it were an inner notebook. In this chapter, we will explore this aspect of mental fictionalism in more detail. Second, the patterns of behaviour that underpin our attributions of mental states often involve the use of external representations, especially language. In Chapter 3 we noted that some mental states seem to be language-dependent: we

---

Mind as Metaphor: A Defence of Mental Fictionalism. Adam Toon, Oxford University Press. © Adam Toon 2023.
DOI: 10.1093/oso/9780198879626.003.0005

cannot display the pattern of behaviour required to possess these mental states without also possessing language. In such cases, believing that p might involve, amongst other things, being disposed to say 'p' under the right circumstances or to write it down. It might also involve using other sorts of representations. For instance, believing that Derby is in the Midlands might involve pointing this out on a map rather than saying so. In this way, our use of material culture can figure in the behavioural patterns required for possessing certain mental states.

Material culture plays other roles in our mental lives besides those we have considered so far. In particular, much of our mental life seems to involve *interacting* with external representations. A canonical example is multiplication with pen and paper. Most of us cannot multiply two three-digit numbers in our heads. The task becomes a lot easier if we're given pen and paper, however. Then we can use the method of long multiplication: we can line the numbers up one underneath the other and work step-by-step through the procedure. By doing so, we transform a complex task (e.g. multiplying 546 by 837) into a series of much simpler tasks (e.g. multiplying 3×4, remembering to carry the 1, and so on). In such cases, we are not simply using an external representation to express a belief (say, that the answer to our puzzle is 457002). Instead, we are interacting with an external representation in order to work out the answer in the first place. We think with the pen and paper.

We use all sorts of material devices in our thinking in all sorts of ways. We use pen and paper to do sums or solve crossword puzzles. We use lists to remember what to buy and diaries to plan our week. We use computers to draw blueprints for buildings or write books about philosophy. In recent years, a growing body of work in cognitive science has shown how much of our cognitive activity has this character, involving cooperation between internal and external resources (for a helpful overview, see Robbins and Aydede 2009). How should we make sense of this? Some authors claim that the extent of our dependence on material culture points towards a radical shift in our conception of the mind. In particular, proponents of the *extended mind thesis* argue that, under certain circumstances, mental states can extend outside our brains and bodies into the world (Clark and Chalmers 1998). On this view, items of material culture, like notebooks, maps, or lists, can become part of the physical basis for our mental states, much like the neurons in our brain (see also Clark 2008; for related ideas, see Rowlands 1999; Wilson 2004; Wheeler 2005; Menary 2007). Let us now turn to consider this idea.

4.1.2 The Extended Mind Thesis

Andy Clark and David Chalmers (1998) introduce the idea of extended mental states through the famous example of Otto and Inga. Inga hears of an exhibition at

86   MIND AS METAPHOR

the Museum of Modern Art (MoMA). She recalls that the museum is on 53rd Street and heads off to see it. In doing so, Inga relies upon her ordinary, biological memory and not upon any external devices. By contrast, Otto is an Alzheimer's patient who suffers from memory loss and therefore carries a notebook with him at all times to record important information. When Otto hears about the exhibition, he looks up the information in his notebook and heads off to see it. Clark and Chalmers argue that, despite their apparent differences, Otto's notebook plays a similar role in his life to the role that Inga's biological memory plays in hers. As a result, they claim, Otto's beliefs are partly constituted by the entries in his notebook. Otto believes that the exhibition is on 53rd Street even before he looks at his notebook, in much the same way that Inga believes this even before consulting her biological memory. The result is that Otto's mind extends beyond his head and body and into the world.

At the heart of Clark and Chalmers' argument is the *parity principle*:

> If, as we confront some task, a part of the world functions as a process which, *were it done in the head*, we would have no hesitation in recognizing as part of the cognitive process, then that part of the world *is* (so we claim) part of the cognitive process.   (1998, 8; emphasis in original)

The parity principle is intended as a heuristic device, encouraging us to form judgements about what counts as cognition and mind behind a 'veil of metabolic ignorance' (Clark 2008, 114). In the case of Otto, it asks us to imagine what we would say if the information about MoMA were located, not in the pages of a notebook, but inside his head. Would we count it among Otto's beliefs? If so, Clark and Chalmers argue, then we ought to say the same about the entry in his notebook. After all, the only difference between the two is the whereabouts of the relevant information. Of course, this does not mean that every time we use a notebook, we acquire an extended belief. Clark and Chalmers stress that Otto's notebook satisfies conditions of 'glue and trust' (Clark 2010b, 83): Otto always has the notebook with him, trusts what it says as a matter of course, and where its information might be relevant, he rarely acts without consulting it (Clark and Chalmers 1998, 17). Most uses of notebooks in everyday life will fail to meet one or more of these conditions. If these conditions are met, however, Clark and Chalmers claim that we have a case of extended belief.

The extended mind thesis has provoked a range of objections and a good deal of debate (for an overview, see Menary 2010). Some have seen it as a radical view that runs counter to our ordinary talk about the mind. Others have argued that it overlooks important differences between biological memory and items of material culture. I shall return to these objections in Section 4.3. First, however, let us consider how ExM relates to representationalism.

4.1.3 ExM and Representationalism

It is important to notice that, despite the controversy it has caused, Clark and Chalmers' argument for ExM fits comfortably with a traditional, representationalist conception of the nature of belief (Crane 2016). Recall that, according to representationalism, having a belief (or desire, hope, fear...) involves having a mental representation which expresses the content of that belief (or desire, hope, fear...). For example, to believe that Derby County will win promotion is to have a mental representation with the content *Derby County will win promotion*. To desire (or hope or fear) that Derby will win promotion involves possessing a mental representation with the same content. The difference between these mental states comes down to the different causal roles that the representation plays in the production of behaviour. Thus, the desire that Derby will win promotion might cause someone to cheer loudly in support, while the fear that they will might lead them to cheer for their opponents instead.

As presented by Clark and Chalmers, the extended mind thesis is entirely compatible with this understanding of the nature of mental states. Indeed, their discussion begins by adopting a broadly representationalist conception of belief in order to argue for the notion of extended mental states: the argument proceeds by pointing to the entries in Otto's notebook and arguing that, since they play a similar role to Inga's biological memory, they ought to count as bearing the content of Otto's beliefs. As Clark and Chalmers put it, '[t]he information in the notebook functions just like the information *constituting* an ordinary non-occurrent belief; it just happens that this information lies beyond the skin' (1998, 13; emphasis added). What the extended mind thesis adds to the standard representationalist story is the idea that the representations that bear the content of our mental states can sometimes be found outside the head. While this addition has certainly proved controversial, it leaves the basic tenets of representationalism intact.

4.2 Minds, Materials, and Metaphors

Fictionalism suggests a different understanding of the relationship between minds and material culture. For the fictionalist, it is not that mental states are typically inner states that sometimes, under special circumstances, extend *outwards* into the world. Instead, much of our talk about the mind is itself a metaphorical projection *inwards* from the world of material culture (Section 4.2.1). As we shall see, however, the fictionalist can still make sense of the notion of extended mental states (Section 4.2.2).

88   MIND AS METAPHOR

4.2.1 Materials as Metaphors

Recall, once again, our re-telling of Sellars' myth. In the first instance, Jones' innovation focused on episodic thoughts—those that occupy our minds from moment to moment. His metaphor for describing such thoughts was based on overt verbal behaviour—asserting things out loud (or denying them, questioning them, and so on). As we noted in Chapter 1, other aspects of our mental lives demand different metaphors. In particular, some are more aptly described by metaphors based on items of material culture such as written, rather than spoken, language. Unlike spoken language, the written word tends to stick around for longer. As a result, it can serve different purposes—both literal and metaphorical. For instance, we can use pen and paper to record information and refer back to it later on. Keeping written records is an important human practice—or, rather, an important set of different practices, from diary writing to bookkeeping. For the fictionalist, the key point is that we can also use these practices as metaphors for understanding our mental lives—principally, of course, our memories. We can treat people *as if* they are able to write down important information inside their heads and refer back to it later on. Likewise, we can use pen and paper to record our wishes or plan our actions, and we can also use these practices as metaphors for understanding our desires or intentions.

From the fictionalist's perspective, then, what is so striking about Otto is that he offers a vivid description of the source for one of our central metaphors for talking about the mind. Thinking of memory as if it were a trusty, ever-present notebook lodged away somewhere inside our heads is a pervasive and powerful way of making sense of people and their behaviour. When someone encounters some important fact or situation in the world, we can imagine them jotting it down in their inner notebook. Later, we can imagine them looking up the information that they've written down and using it to guide their actions. Indeed, the rules that we suppose to govern the use of this inner notebook are summed up fairly well in Clark and Chalmers' description of Otto, especially the 'glue and trust' conditions: our inner notebook is always present, fairly trustworthy, and we always consult it when we need it. In this way, the fictionalist suggests, we use our practices for using external representations—in this case, notebooks—to make sense of people and their behaviour. (As we have noted already, the metaphor of memory as a notebook can also lead us astray. We will consider this point in Section 4.4. See also Draaisma 2000.)

According to the fictionalist, the metaphor of memory as a notebook acts to fill out one region of our folk psychological discourse. Looking at Italy as if it were a boot yields the rules of the Italy-as-a-boot game, allowing us to make assertions about the coastline of Italy by making utterances within this game. In a similar manner, looking at people as if they possessed an inner notebook yields the rules of the game for attributing standing beliefs. If we say 'Inga believes that MoMA is on 53rd Street', we are not claiming that she has an inner representation with this content; we are indicating how to pretend correctly within our game. If we had said, 'Inga believes that MoMA is on 52nd Street', our pretence would have been inappropriate. Why? Because of a whole range of facts about Inga's behaviour: when she hears about the exhibition, she heads off to 53rd Street, not 52nd Street; when someone asks her for directions to MoMA, she says '53rd Street', not '52nd Street'; and so on. The metaphor of memory as a notebook gives us a way of picking out this complex pattern in Inga's behaviour, just as the metaphor of Italy as a boot gives us a way of describing its coastline.

How does all this relate to the extended mind thesis? We saw in Section 4.1 that Clark and Chalmers' argument for ExM begins from a representationalist account of belief. Beliefs are taken to be mental representations with particular causal roles. What ExM adds to this basic vision of the mind is the idea that these representations can sometimes be said to extend outwards into the world. In a sense, the fictionalist approach is precisely the reverse. For the fictionalist, the mind itself is a metaphorical projection *inwards* from the world of material culture. In many respects, it is external representations that come first. In particular, we draw on cases in which people use external representations (e.g. recording information in notebooks) to talk about cases in which they do not (e.g. relying on their biological memory). The question remains, however: what should the fictionalist say about cases in which people actually *do* rely upon external representations, like notebooks, blueprints, and shopping lists? Can fictionalism make sense of the idea of extended mental states? In what follows, my aim will be to show that it can and that, in fact, fictionalism provides an alternative way to understand ExM that can avoid many of the objections levelled against it.

4.2.2 Materials as Minds

Can fictionalism make sense of the idea of extended mental states? For the fictionalist, terms like 'belief' and 'desire' are fundamentally metaphorical. So our question becomes: in what cases should we apply these metaphors? In what cases are such metaphors apt or useful? In particular, should we apply them only in cases where people rely on their brains and bodies alone—or are they also useful when people are dependent on external devices? At first glance, this might seem like an odd question. After all, the fictionalist claims that the world of external representations is the *source* of our metaphors for describing the mind. Using the world of external representations as a metaphor for talking about the world of external representations would not seem terribly helpful: we seem to be asked to apply the metaphor to its own source. Asking someone to see Italy as a boot might be a helpful way of describing its coastline, but asking them to see a boot as a boot is not especially illuminating.

90 MIND AS METAPHOR

When we look more closely, however, we see that fictionalism is able to make sense of the notion of extended mental states. For often we use one sort of external representation to make sense of another. Consider memory and belief. According to the fictionalist, our talk about memory is guided by a core metaphor of memory as an ever-present and trusted notebook. We can use this metaphor to make sense of how people behave without actual notebooks to hand. And yet we might also apply it when people rely on other tools and external representations. Suppose that, rather than writing down addresses, Otto drew little maps and sketches in his notebook to help him find his way around. We might still say that Otto believes that MoMA is on 53rd Street, even if this is not written down anywhere in his notebook. What we are doing in this case, according to the fictionalist, is using our understanding of one form of representation to make sense of another. We are saying that, for some purposes, we won't go too far wrong if we treat Otto as if he had a notebook with the address written down. Saying this does not depend upon controversial claims about being able to reduce the content of maps or pictures to propositions. It simply relies on the fact that people can *do* similar things with maps and pictures as they can with written text.

Similar points apply in the case of extended desires or extended intentions. In Chapter 1, I suggested that we think of someone's desires as if they were a kind of inner shopping list (or 'wish list'). We can use this metaphor to make sense of how people behave without actual shopping lists to hand. For instance, we might make sense of my behaviour one afternoon (e.g. walking to a café on campus, standing at the till, saying the words 'tea and a packet of biscuits, please') by pretending that my actions are guided by a private, inner shopping list that expresses my desires (e.g. 'cup of tea', 'biscuits', 'nice sit down', etc.). And yet we can also use this metaphor to make sense of how people behave when they rely on tools and external representations. For instance, I might take photos of items I would like for my birthday, rather than writing a wish list. My family might still remark that I'd like, say, a blue rucksack, even if I never write this down anywhere—or commit it to (internal) memory. The same is true of intentions. I suggested that we think of someone's intentions as if they were a kind of inner plan or itinerary. Again, this metaphor can be useful for making sense of people's behaviour when they are relying on their bare brains and bodies—but also when they rely on external devices too. Consider an architect's blueprints. Looking at her drawings, we might note that the architect intends to build an imposing three-storey mock-Georgian house with a symmetrical front portico supported by Doric columns. This description of the architect's intentions might be perfectly correct, even if none of these features are explicitly stated anywhere in the plans. (In Chapter 5, I will offer a similar account for concepts, knowledge, and understanding.)

The picture that emerges is as follows. Humans have a remarkable capacity to use all manner of different external, material representations. This is an enormously important feature of human life in its own right, of course. According to the fictionalist, it also shapes our vision of the human mind as an inner realm. For some of the most important ways in which we use external representations—for recording information, stating our wishes, for planning our actions—also lie at the heart of our language for talking about the mind. We use these external devices as the source of metaphors for making sense of people and their behaviour. It is these metaphors that give meaning to our attributions of mental states like belief, desire, and intention. And yet, when we come to apply these metaphors, we do not restrict ourselves to human activity in the absence of tools and external representations. When we attribute states like belief and desire, our main concern is to make sense of people's behaviour. Often, that behaviour is itself dependent upon interaction with certain sorts of tools and external representations—sometimes ones that are quite different from the original source of our metaphors.

Fictionalism yields a rather different set of criteria for attributing extended mental states to the usual, representationalist interpretation of ExM. Recall that, according to representationalism, beliefs are mental representations that play a characteristic causal role. When we attribute a belief to someone, we are claiming that their cognitive machinery contains such a representation. Clark and Chalmers' strategy is to show that, in some cases—such as Otto's—representations that play the appropriate causal role can be found outside the head. On this view, to judge whether we are dealing with a genuine case of an extended mental state, we should ask ourselves: does this external representation play the right causal role to count as a belief (or desire, or intention, etc.)? Since our intuitions are taken to be clearest when applied to inner states, the parity principle suggests that we imagine that the external representation was instead found internally before delivering our judgement.

In contrast, fictionalism suggests that even attributions of ordinary, non-extended mental states involve an imaginative move akin to that proposed by the parity principle: we make sense of people by imagining that they had inner analogues of external representations like notebooks (or lists, or itineraries, etc.). When judging whether we are faced with a case of an *extended* mental state, we should no longer focus on a particular external representation and ask ourselves whether it plays the appropriate causal role to count as a belief. Instead, we should ask: does the person using this external device display the appropriate overall pattern of behaviour for us to apply our stock of folk psychological metaphors? And when we answer this question we rely not upon a prior grasp of beliefs as inner states, but on our understanding of the way people use external, public representations, like notebooks.

4.3 Defending ExM

Let us now see how reinterpreting the notion of extended mental states along fictionalist lines can help us to respond to some well-known objections to ExM. In each case, I will suggest, the root of the problem lies not with ExM

92 MIND AS METAPHOR

but with representationalism—and turning instead to fictionalism shows us a way out of trouble.

4.3.1 Common Sense

The first, and most straightforward, objection to the extended mind thesis is simply that it clashes with our common-sense view of the mind. Clark and Chalmers try to motivate ExM by appealing to our ordinary, folk psychological conception of belief. And yet, it is argued, we ordinarily think of mental states as inside the head. The folk are internalists. Talking of beliefs extending into notebooks therefore stands radically at odds with our common-sense view of the mind (e.g. Adams and Aizawa 2001; Rupert 2004).

This objection is less troubling than it might seem at first sight. When we consider ordinary talk about the mind more closely, I think, we see that we often attribute extended mental states. David Houghton (1997) offers a number of examples. Suppose that, before heading out to the supermarket, Ted inspects his kitchen cupboards and writes down a list of items he wants to buy: a pint of milk, bread, cornflakes, carrots, and so on. It would be quite natural to say that Ted wants to buy, say, milk and carrots, even if he's unable to recall these items without his list to hand. After all, this is precisely the reason that we use shopping lists. Or consider an architect's technical drawings. It would be entirely in line with our ordinary way of talking to say that these drawings capture the architect's intentions for the building. Saying this does not rely upon assuming that the architect must have committed every aspect of her drawings to memory. Indeed, if the drawings are sufficiently complex, it might not even be possible for the architect to do this. And yet we should still say that she intends to construct such-and-such a building.

So there are reasons to think that ExM might not depart so far from our ordinary talk about the mind after all. In fact, Houghton argues, it is internalism that is at odds with ordinary usage. Many remain unconvinced, however—even those who are otherwise sympathetic to ExM. Thus, Mike Wheeler imagines asking the folk a question that directly concerns the whereabouts of a particular cognitive state, such as '"Where in space are the relevant cognitive states of the architect realized?"' (2011, 424). Not implausibly, Wheeler suggests that the likely response is an internalist one. So it seems that folk opinion is rather conflicted: on the one hand, we commonly attribute extended states; on the other hand, we balk when asked about the location of those states.

Fictionalism can explain these apparently conflicting responses to ExM. The fictionalist claims that all attributions of mental states involve metaphor: we talk *as if* people were guided by representations that capture the content of those states. And it is the metaphorical nature of mental state attribution, I believe, that explains why our intuitions falter when asked explicitly about the whereabouts of mental states. For, properly speaking, mental states are nowhere. As we saw in Chapter 1, to ask where beliefs *are* is to ask a 'silly' question—one that pushes a metaphor too far (Yablo 1998; cf. Di Paolo 2009). It is rather like being told that someone has a chip on his shoulder and asking if it is on his right or left shoulder. This feature of our ordinary talk about the mind is not confined to cases of extended mental states. To ask whether beliefs lie inside or outside the head is, I think, no more (or less) silly than asking whether they lie on the left or right side of the skull. Both questions misunderstand the nature of our talk about mental states. Both push a metaphor too far.

4.3.2 Intentionality

A rather different line of criticism levelled against ExM has been to allege that it ignores the crucial distinction between the intentionality of mental states and that of public representations, such as spoken and written language. As we saw in Chapter 3, it is widely held that the intentionality of public representations is derived from the intentionality of mental states. Furthermore, according to representationalism, the intentionality of mental states is, in turn, to be explained in terms of the content of mental representations. In contrast to public representations, mental representations are said to possess original intentionality: their content does not depend upon any other intentional states. Instead, it must be explained in non-intentional terms, such as causal relations or biological function. From this perspective, critics have argued that the entries in Otto's notebook cannot count as his beliefs, since they possess merely derived, rather than original, intentionality: their content depends upon social conventions and the mental states of those who use them (e.g. Adams and Aizawa 2001, 2008; Fodor 2009).

We saw in Chapter 3 that fictionalism rejects this approach to intentionality. To say that someone has a certain mental state is not to say that they possess a particular inner representation; it is to say that they behave *as if* they had such a representation. The intentionality of mental states is grounded not in the content of any inner representations, but in facts about a person's behaviour: possessing a given mental state depends upon exhibiting the right pattern of behaviour. Moreover, exhibiting this pattern of behaviour will typically involve engaging in the use of public representations, especially language. For example, one reason that it is appropriate to say that Inga believes that MoMA is on 53rd Street is that, if you ask her where MoMA is, she'll answer '53rd Street'. In this way, the fictionalist claims, the intentionality of mental states is derived from the intentionality of public representations, rather than vice versa.

In the present context, the important point is that, by adopting the fictionalist's approach to intentionality, ExM can avoid the objection that it ignores the

94   MIND AS METAPHOR

distinction between the intentionality of mental states and public representations. According to the fictionalist, what happens in cases of extended mental states is *not* that public representations are somehow illegitimately co-opted to serve as mental representations. Instead, what is distinctive of such cases is that, by interacting with public representations or other external objects, someone manages to exhibit the right overall pattern of behaviour to count as possessing an extended mental state. Thus, by interacting with his notebook, Otto is able to exhibit broadly the same pattern of behaviour as Inga—he too can reply '53rd Street' when you ask him MoMA's address, for example. In this way, cases of extended mental states can be, at one and the same time, instances of *both* forms of intentionality: they are cases in which the intentionality of public representations and the intentionality of mental states can overlap (cf. Adams and Aizawa 2008, 38).

Notice, however, that although these two forms of intentionality *can* overlap, they need not do so. In this respect, Otto's case is potentially misleading. In Otto's case, the content of his notebook entry and the content of his belief are the same. In other cases, however, they can come apart. To see this, notice that, for the fictionalist, Otto counts as having the belief that MoMA is on 53rd Street *not*, strictly speaking, in virtue of the fact that his notebook contains a representation with this content; instead, it is in virtue of his overall pattern of behaviour. In principle, Otto might exhibit the same pattern of behaviour even if the content of his notebook entries were different. For example, suppose that Otto realizes he has made a mistake when originally writing down addresses in Manhattan: he counted the street numbers wrongly and they're all one number too low. If he's feeling lazy, he might not bother updating his notebook but just remember to add one every time he looks them up. In this case, the content of his belief might remain the same (he believes that MoMA is on 53rd Street) while the content of his notebook entry is different (it now says 'MoMA is on 52nd Street').

More dramatically, we might even imagine cases in which an external device bears no representational content at all *qua* external object and yet still enables someone to possess extended mental states. Suppose that James suffers from anxiety but finds that using a stress ball helps him to remain calm (cf. Colombetti and Roberts 2015). Without his stress ball, let us assume, James finds himself rather overwhelmed at work: he fears that tasks will never be completed, worries that colleagues doubt his abilities, wishes he could avoid responsibilities, plans to find a new job, and so on. Thankfully, with his stress ball to hand, he finds matters altogether more manageable. As a result, we might attribute all manner of more positive beliefs, desires, and intentions to James: he now believes that the troublesome matter with his client can be resolved, that his boss values his contribution to the team, wishes to handle the new contract himself once it is awarded, plans to build his career at the company, and so on. Like Otto's notebook, the stress ball allows James to possess extended mental states. But it achieves this feat not by expressing the content of those states, but by helping to regulate and guide James's overall behaviour in such a way that our attributions are appropriate.

4.3.3 Cognitive Science

As well as questioning ExM's credentials vis-à-vis common sense, critics have also worried about its implications for cognitive science. The main concern here is that extended cognitive systems will simply be too heterogeneous to form a proper basis for scientific investigation. Thus, Adams and Aizawa (2001, 61) ask us to:

> Consider...the range of tools humans use as mnemonic aids. There are photo albums, Rolodexes, computer databases, strings around the finger, address books, sets of business cards, bulletin boards, date books, personal information managing software, palmtop computers, hand drawn maps, and lists of 'things to do'. What are the chances of there being interesting regularities that cover humans interacting with all these sorts of tools? Slim to none, we speculate. There just isn't going to be a science covering the motley collection of 'memory' processes found in human tool use.

Clark (2010a) considers two main ways of responding to this worry. The first, more optimistic, response points out that even apparently dissimilar causal mechanisms might turn out to display regularities when seen from a suitable overarching framework, such as that of 'information storage, transformation and retrieval' (2010a, 50). The second, more pessimistic, response that Clark considers (but does not endorse) takes the real lesson of ExM to be that 'the realm of the mental is itself too disunified to count as a scientific kind' (2010a, 62) and that perhaps we ought therefore to *'eliminate the mind'* (2010a, 63; emphasis in original).

Fictionalism allows us to avoid the threat of eliminativism for mental states. We seem to be faced with a dilemma: either terms like 'belief' and 'desire' pick out some underlying state or processes that form a scientific kind, or else they must be eliminated. But this is a false dilemma. After all, there are many other legitimate uses of language, even in scientific contexts. According to fictionalism, terms like 'belief' and 'desire' function as metaphors that allow us to pick out certain patterns in behaviour. What unites cases of extended belief is that they each exhibit this pattern of behaviour, not that they share any common inner state or process. As a result, it is no argument against ExM—at least if we adopt a fictionalist approach—to point out that cases of extended belief (or desire, or intention, etc.) might contain an unruly motley of different internal processes.

Here we might recall Dennett's well-known example of Jacques, the Frenchman who kills his uncle in Trafalgar Square (Dennett 1987, 54). Jacques is caught by

96 MIND AS METAPHOR

Sherlock, while Tom reads about the crime in the *Guardian* and Boris in *Pravda*. All four share a belief: they all believe that a Frenchman has committed murder in Trafalgar Square. And yet, Dennett submits, it is hard to see why they must all share a similarly structured object inside their heads. The fictionalist shares this intuition for ExM cases too. Consider James again, who uses a stress ball to alleviate anxiety. Now consider his colleagues Jess, John, and Jane. Jess achieves a similar boost to her mood by relying on a book of self-help mantras, while John relies instead on anti-depressants. Finally, Jane is more naturally relaxed and confident. All four might share the belief that the troublesome difficulty with their client will be resolved. And yet it is hard to see why they must all share a similarly structured object either inside or outside their heads.

So the fictionalist need not worry that cases of extended mental states are unlikely to show much unity in their underlying states or processes. At the same time, fictionalism can explain why we might be misled to expect such unity. For, according to the fictionalist, we make sense of *all* cases of belief—however diverse their underlying machinery, whether extended or not—by treating them *as if* they were guided by inner sentences that play a particular role. If we begin to take such talk too seriously, then we will be tempted to misconstrue terms like 'belief' and 'desire' as attempts to pick out real inner states or processes with distinctive characteristics, such as possessing a certain sort of content or playing a particular causal role. Seen correctly, however, our ordinary talk about the mind does not carry such implications.

Of course, even if the existence of a motley of underlying processes does not threaten the idea of extended mental states, we might still worry about its implications for cognitive science. Would any attempt to construct a science of such a diverse assortment of systems, states, and processes be doomed at the outset? It is hard to see why. We must be careful not to hanker after an unrealistic ideal of the unity of science. After all, even the principles of Newtonian mechanics are, strictly speaking, true of few real systems. Instead, Newtonian mechanics offers a set of models (free fall, the ideal oscillator, and so on) that apply more-or-less accurately in a diverse range of areas, from falling leaves to fluid flows (e.g. Giere 1988). It is not clear why a science of extended cognition might not aim for a similar unity. Moreover, even if this aim proves unrealistic in the long run and cognitive science turns out to be rather less unified than Newtonian mechanics, this would hardly be especially damning. Most sciences must invoke a range of different principles and models to encompass their subject matter (Clark 2010a).

4.3.4 Drawing Boundaries

We can now consider a final line of criticism concerning ExM. As well as ignoring a crucial distinction concerning intentionality, ExM has also been charged with overlooking important differences in the causal profiles of internal and external processes. In this vein, a number of critics have drawn attention to differences between biological memory and external devices like Otto's notebook (Adams and Aizawa 2001; Rupert 2004; Weiskopf2008). For example, psychologists have found that ordinary, biological memory exhibits a phenomenon called *negative transfer*, whereby old memories interfere with the ability to form new ones. If I have memorized your old phone number, for instance, it will be harder for me to memorize your new number than it would be if I'd never known the old one in the first place. When using his notebook, it seems that Otto won't exhibit negative transfer: he can just rub out the old phone number and write down the new one. Similar points can be made for other important characteristics of biological memory that psychologists have described, such as so-called generation effects or power laws of remembering and forgetting (Rupert 2004; Sprevak 2009). Given such differences, critics argue, the entries in Otto's notebook do not count as beliefs since it is simply false to say that they play the same causal role as items in Inga's biological memory.

In response, proponents of ExM have justly questioned whether these features of biological memory are essential to belief. Recall that Clark and Chalmers' claim is that the notebook entries play the causal role *that folk psychology attributes to beliefs*. It is surely no part of the folk notion of belief that people's memory exhibits effects like negative transfer. After all, describing such effects has taken detailed scientific investigation. This point is often expressed in terms of the 'grain' in which we ought to specific the relevant causal roles. Effects like negative transfer are said to be too fine-grained to appear in the folk notion of belief. If we adopt a more coarse-grained specification of the causal role of belief, then Otto's notebook can count as part of his mind (Sprevak 2009; Clark 2010a). Unfortunately, though, this response threatens to generate trouble in the opposite direction. For if the relevant causal roles are specified too coarsely, critics foresee a rampant expansion of mind into the world. If I am often sat at my computer, do the contents of its hard drive count as my beliefs? What about the books on my office bookshelf or in the university library? Or the entire contents of the internet? Proponents of ExM would therefore seem to face a serious challenge when it comes to demarcating the boundaries of the mind. How can they specify the causal role for belief so as to allow for some cases of extended beliefs (like Otto) while avoiding a dramatic expansion of mind into the world? (Sprevak 2009; Farkas 2012; Crane 2016).

Fictionalism offers a way to resolve this dilemma, since it rejects the idea that beliefs are individuated by their causal role. At the heart of our talk about standing beliefs, the fictionalist insists, is not a definition of a causal role, but a metaphor— that of memory as a trusty, inner notebook. This metaphor can be apt and useful when applied to ordinary, biological memory, even if our biological memory exhibits effects that are absent from notebooks, like negative transfer. After all,

98 MIND AS METAPHOR

this is a characteristic feature of metaphors: we use metaphors to relate a primary domain (e.g. memory, Italy) with a secondary domain (notebooks, boots), not to claim that they are exactly alike. The same lesson applies when we consider extended beliefs. The metaphor of memory as a notebook can be useful for understanding our interactions with external devices even when they function rather differently to written entries in notebooks, like pictures, maps, or even smartphones. At this point, of course, critics will again see the risk of a dramatic expansion of mind into the world. How are we to draw a line around genuine cases of extended belief?

The correct response to this worry, I think, is to insist that there is no sharp line to be drawn. We've already noted on a number of occasions that the aptness of metaphors is not an all-or-nothing affair. In many instances, everyone will agree whether a metaphor is apt or not. In borderline cases, however, there is room for disagreement. Are the clouds angry—or has the sun just gone in for a moment? Is Ruth standing at a crossroads—or is the decision she faces fairly trivial in the grand scheme of things? As we saw in Chapter 1, the same lesson applies to talk about the mind. In some cases, there will be agreement over the best way to describe a person's mental state. In other cases, competent speakers may differ. Indeed, even an individual speaker may feel uncertain if asked whether the metaphor is appropriate. Consider the numbers stored on a mobile phone. Mine includes the number of Exeter Library (01392 407027), but I can't recall this without using my phone. Do I believe that Exeter Library's phone number is 01392 407027? Well, yes and no. In many respects, the familiar metaphor of memory as a notebook captures my behaviour fairly well: if I need to call the library, I will call 01392 407027; if you ask me the area code, then I'll reply '01392'; and so on. In other respects, the metaphor is less apt: it'll probably take me a while to tell you the area code, as I'll need to look it up in my phone; at times when my battery is dead, I won't be able to tell you; and so on.

The key point is that, for the fictionalist, the boundaries of the mental are somewhat blurred: there are borderline cases in which there are aspects of someone's behaviour that our metaphors capture well and other aspects that it does not. Interestingly, Clark and Chalmers themselves acknowledge that there may not be 'categorical answers' to give regarding all putative cases of extended belief (1998, 17). Of course, this will seem unsatisfying to those who want to draw a clear boundary between the mental and non-mental (cf. Sprevak 2009, 522–3). And yet, I believe, this feature of fictionalism fits well with our ordinary talk about the mind, which is similarly ambiguous and conflicted. Consider the library phone number again. Suppose I'm late to give a talk at the library. In this context, the events manager might say, 'I wonder why he hasn't called us—he knows the number.' On the other hand, when I arrive I might offer an excuse: 'I'm sorry I couldn't call you—my phone was dead and I didn't know the number.' In these contexts, I submit, neither of these ways of speaking strikes us as especially odd or out of place. Both simply stress different aspects of a metaphor for different purposes.[1]

Once again, notice that this is not a phenomenon unique to extended mental states. Even in cases where we do not rely on external devices, our talk about the mind can be equivocal in much the same way. Suppose that I didn't have Exeter Library's number stored in my phone, but still exhibited a similar pattern of behaviour: sometimes, with some effort and delay, I'm able to tell you the number; at other times, if I'm feeling tired or distracted, I struggle to recall it (cf. Houghton 1997; Sprevak 2009; Farkas 2012). In this case too, I suggest, we would also hesitate and feel conflicted when asked if I knew the phone number (e.g. 'Come on, you know this!' versus 'Come on, you really ought to know this by now!'). As we saw in Chapter 3, a similar ambiguity arises when we are asked about the beliefs of animals or pre-linguistic infants. Does Rover believe he lives at 99 Canine Avenue? Well, yes and no. Treating Rover as if he had a trusty inner notebook with his address written down will help to make sense of some aspects of his behaviour (e.g. that he heads back to 99 Canine Avenue if he gets off his lead) but not others (e.g. if you ask him his address, he's not especially forthcoming).

4.4 Folk Miracles?

We can now return to an important challenge for mental fictionalism, namely that it makes the success of folk psychology a miracle. In Chapter 3 we saw that representationalists give two arguments for the existence of mental representations that are familiar from debates over scientific realism. The first argument is an inference to the best explanation: we should infer that folk psychology is true since it is the best explanation of people's behaviour. The second argument is the no miracles argument: we should infer that folk psychology is true since its truth is the best explanation of its success (e.g. its success in predicting people's behaviour). If folk psychology weren't true, then its success would be a miracle. I claimed that both of these arguments fail because the story that folk psychology tells about our inner machinery invokes real fictions—claims that are not merely false, but incoherent—and so simply cannot be true. This blocks the standard realist inference from explanatory success to truth, at least in the case of folk psychology. It also means that we cannot appeal to the truth of folk psychology to explain its predictive success. And yet it now seems as if we are left with a mystery: if the truth of folk psychology does not explain its success, then what does? Does

---

[1] In Chapter 5, I'll suggest that, in fact, the metaphor that underpins our talk about knowledge is different to that involved in talk about belief. For the moment, all that matters is that both *are* metaphors, so that their application is sometimes ambiguous or conflicted.

100 MIND AS METAPHOR

the fictionalist have to settle for the idea that the success of folk psychology is an inexplicable miracle?

One worry concerns the idea that the metaphors that underlie our ordinary talk about the mind might be indispensable. In Chapter 2, I suggested that the metaphors of folk psychology are representationally essential: they cannot be given a literal translation into language referring solely to behaviour. Daniel Hutto (2022a) sees a dilemma here for the fictionalist. If the metaphors of folk psychology *could* be reduced to, say, talk about behaviour or neurons, then we would be able to explain their usefulness. But then these metaphors would be merely a practical expedient, and mental fictionalism would lose much of its appeal. Alternatively, if the fictionalist insists that our metaphors *cannot* be reduced in this manner, she will be unable to explain their usefulness. Folk psychology will remain in glorious (or, rather, inglorious) isolation from the rest of cognitive science, and the fictionalist must remain content with 'taking it as a brute, inexplicable fact that pretending to ascribe [folk psychological] states to certain things is useful even though we can never give a deeper explanation of why this is the case' (2022, 173). In short, fictionalism will leave us with a 'perpetual mystery' (2022, 173).

How might the fictionalist respond to this dilemma? Broadly speaking, I think the correct response is to insist that the demand for reduction is too strong, and again rests upon an unrealistic ideal of the unity of science. It is true, of course, that there are some notable cases in which scientists have been able to reduce a 'higher' level theory to a 'lower' level one. A classic (though still contested) example is the reduction of thermodynamics to statistical mechanics, which enables us to explain, for instance, the laws relating the temperature and pressure of a gas in terms of those governing the behaviour of its molecules. This is a remarkable achievement. And yet it is important to remember that cases like these are relatively few and far between. In fact, when we look across the sciences, we find many 'higher' level theories (e.g. in chemistry or biology) that are successful, but which have not been (and, arguably, never will be) reduced to any 'lower' level, or more fundamental theory (e.g. in physics). From this perspective, the demand for reduction seems too strong, since it would imply that the success of many of our higher level theories is equally mysterious.

Even if the demand for reduction is too strong, however, the fictionalist still seems to face difficulties here. The trouble seems to arise, not so much from her claim that folk psychology is indispensable, but from her claim that it is false. To see why, notice that a realist about folk psychology—someone who thinks it provides a true theory of our inner machinery—has a ready explanation for its success, even if she denies that it can be reduced to some other, more fundamental, theory. (Indeed, Fodor 1974 endorses a view of this sort.) The realist argues that folk psychology is successful because it is true. This explanation relies only on the fact that a true theory gives true predictions. For this purpose, it does not

MINDS AND MATERIALS   101

matter whether that theory can ultimately be reduced to some other, more fundamental, theory. Of course, if folk psychology cannot be reduced in this way, we might be left without any deeper explanation for why its laws hold, in the way in which we can explain why the laws of thermodynamics hold by appealing to statistical mechanics. In this sense, folk psychology might remain mysterious. Still, its *success* would be no mystery: it is no mystery that the laws of folk psychology enable us to make true predictions if those laws are true. And yet the fictionalist cannot offer this explanation: she cannot explain the success of folk psychology by appealing to its truth, since she thinks that it is not true. *Can fictionalism explain the success of folk psychology—or must she regard it as a miracle?*

We can begin to dispel the sense of mystery by considering extended mental states—that is, states in which people rely on tools and external representations. For the fictionalist, the success of folk psychology in dealing with these states is far from miraculous. It is hardly a miracle that we are successful if we treat Otto as if he had a trusty, ever present notebook to write down useful information. After all, that is exactly what he does have! In Otto's case, the story told by folk psychology is true. Clearly, this does nothing to establish the existence of mental representations. A similar lesson holds even in more mundane cases of extended mental states. Consider a well-known example given by Fodor:

> Someone I don't know phones me at my office in New York from—as it might be—Arizona. 'Would you like to lecture here next Tuesday?' are the words that he utters. 'Yes, thank you. I'll be at your airport on the 3p.m. flight' are the words that I reply. That's *all* that happens, but it's more than enough; the rest of the burden of predicting behaviour—of bridging the gap between utterances and actions—is routinely taken up by theory. And the theory works so well that several days later (or weeks later, or months later, or years later; you can vary the example to taste) and several thousand miles away, there I am at the airport, and there he is to meet me. Or if I *don't* turn up, it's less likely that the theory has failed than that something went wrong with the airline.
>
> (1987, 3; emphasis in original)

It might well be that Fodor himself was able to remember his flight time to Arizona a week (or a month or a year) in advance. But I take it that most of us (me included) would need to rely on external tools to achieve this: we would probably write down the details in our diary, make a note on our calendar, or put a reminder in our phone. Nowadays, we would probably also receive umpteen messages from the airline too, reminding us of our flight details and trying to sell us car hire. The key point is that, to the extent that we do rely on such external devices and prompts, the success of our prediction looks far less miraculous, at least by fictionalist lights. In the end, all it boils down to is that, in many respects,

102 MIND AS METAPHOR

someone using a diary (or calendar, or phone reminders, etc.) behaves in a similar manner to someone using a notebook.

Now consider mental states that are not extended—that is, mental states in which people do not rely on external tools or representations, but only their brains and bodies. The first point to notice is that in such cases the success of folk psychology is rather less impressive. As a means for understanding biological memory, for example, the metaphor of memory as a notebook is extremely useful, but it is far from perfect. We are all too well aware that if people cannot write down important information, they often get muddled up, misremember things, or forget them entirely. That is why we use things like notebooks, diaries, and calendars. Consider Fodor's example again. Imagine that you were invited to Arizona to give a lecture months or years in advance but had no access to a diary (or calendar, or phone, etc.) to write down the details and no reminders from the airline to help you. At least in my own case, I am willing to confess that if I were to fail to turn up at the airport in Arizona at the allotted time, the likeliest explanation would be simply that I had got confused or forgotten my flight details, not that something had gone wrong with the airline.

It is important to remember, then, that our folk psychological predictions are not always successful. Despite our best efforts, we are often wrong about what a person will do in the future, even when the person in question is ourselves. It is also important to notice that, even when our prediction of what a person will do is correct, it is often because they have modified their behaviour in order to *make* our prediction correct. Once again, Fodor's example is instructive. If I were to agree to give a lecture in Arizona, I would not only be taking on board some information that I need to remember; I would be making a commitment to the person who invited me. In particular, I would be making a commitment to behave in a certain way (e.g. to turn up at the airport in Arizona at a certain time). This commitment will shape my future behaviour: since I know I am expected to fit into a certain pattern of behaviour, I do everything I can to make sure that I do fit into this pattern. For instance, I write down the details in my diary. Or, if I don't have access to my diary, I might make a concerted effort to keep reminding myself of the details instead. In such cases, it would seem, it is not so much that folk psychology fits our behaviour; instead, we modify our behaviour to fit folk psychology.

This point is often put by saying that the norms of folk psychology *regulate* our behaviour (McGeer 2007). This regulative aspect of folk psychology might seem at odds with a fictionalist approach (Hutto 2022a). As far as I can see, however, the fictionalist can happily acknowledge that folk psychology regulates our behaviour, as well as helping us to explain and predict it. In fact, this point follows naturally from the analogy with games of make-believe. Consider children playing with a doll. The rules that govern this game tell the children what to imagine about their behaviour. For instance, the rules mean that raising a cup to the doll counts as feeding the baby or that rocking it in your arms counts as rocking the baby to sleep. Notice, however, that the rules of the game also regulate the children's behaviour as well. For instance, the children raise the cup to the doll *because* this counts as fictionally feeding the baby and rock the doll in their arms *because* this counts as rocking the baby to sleep, and so on. In this way, the rules that govern games of make-believe not only shape the participants' imaginings; they also shape their behaviour. In fact, people playing games tend to behave in fairly regular ways. They do so precisely because those ways of behaving fit with the rules of the game.

So, the predictions of folk psychology are not always true and, when they are true, their truth is often self-fulfilling. This might help to dispel some of the mystery surrounding the success of folk psychology. Still, folk psychology does make successful predictions some of the time, and it is doubtful whether its success can always be explained by its regulative dimension. Can fictionalism explain this (more limited sort of) success? Or must we rest content with the idea that it is an unexplained miracle—a 'cosmic coincidence' (Smart 1963)? At this point, it is helpful to distinguish between *accommodation* and *prediction*. In the case of accommodation, we construct a theory to fit empirical evidence that is already available to us. In the case of prediction, we first construct a theory and use it to formulate an empirical claim; only afterwards do we check whether this empirical claim is true. Realists are typically more impressed by successful predictions than by accommodations. It is predictive success that seems to cry out for an explanation. Why is this? Lipton (2005) suggests that the difference is that, in the case of accommodation, there is an alternative explanation for why a theory fits the facts: we have 'fudged' it. In other words, we have (perhaps unwittingly) constructed an ad hoc theory that fits the facts available, but is likely to have few other theoretical or empirical virtues. Like the Ptolemaic astronomer, for example, we might have managed to get our theory fit the data only by adding epicycle upon epicycle, sacrificing its theoretical elegance and simplicity as a result. It is not a miracle that our theory accommodates the facts: we built it to do exactly that— and, in doing so, we created a theory that has few other virtues and is unlikely to be true.

How does all this relate to folk psychology? It is clear that folk psychology has managed to accommodate a wide range of facts about human behaviour. It is less clear, I think, that folk psychology can claim any genuinely novel predictive success. This is a complaint that is often made by eliminativists, of course. Paul Churchland (1981) notes that, although folk psychology is successful within its own domain, it says little about less familiar aspects of our mental life, such as sleep or mental illness. Where folk psychology has had (limited) success in such areas, it is often because it has been further refined to accommodate the facts available (e.g. by introducing the notion of unconscious beliefs). For the fictionalist, none of this counts against folk psychology, since it is not intended as a theory of an inner realm. It conjures up an inner realm of representations not as a theory, but as a metaphor. The point is simply that, when the representationalist challenges us to explain why talking in this way works, the fictionalist can offer an explanation. Folk psychology works because we have fudged it: over many years, we have purpose built a vast and rather haphazard collection of different metaphors that accommodate known patterns in human behaviour. This is a remarkable achievement and we would be lost with it. But it is no miracle.

4.5 Conclusion

Mind and material culture are deeply intertwined. Material culture furnishes the rich warehouse of metaphors that make up our concept of mind. And these metaphors, in turn, pick out patterns of behaviour that often involve the use of material culture. When we look at these patterns of behaviour, material culture sometimes seems to play a fairly passive role: we use pen and paper simply to say what we think. In this chapter, however, we have seen that material culture can play a more active role in our mental lives: by interacting with pen and paper (or computers, or smartphones, or even stress balls) we manage to display new patterns of behaviour and thereby possess new mental states. In Chapter 5, we will see how this approach can help us to understand the nature of inquiry.

5

Inquiring Minds

Our vision of the mind shapes our vision of the nature of human knowledge. Many of our most influential thinkers about knowledge—such as Descartes, Locke, Berkeley, and Hume—have treated the mind as an inner world of representations. This has influenced their vision for epistemology: its central questions concern the nature of our inner world and its relationship to the world outside—if, indeed, there *is* a world outside the mind. In this chapter, my aim will be to show that mental fictionalism offers a new perspective on the nature of inquiry. After looking briefly at some existing views (Section 5.1), I will apply the framework developed in earlier chapters to three aspects of the mind that are central to inquiry: concepts, knowledge, and understanding (Section 5.2). For the fictionalist, epistemology is not an attempt to understand the workings of an inner world, forever screened off from the world outside. Instead, it is an attempt to map out the complex array of interactions that take place between people and tools. I will argue that this vision for epistemology has important implications for our understanding of the nature of inquiry, its history, and its limits (Section 5.3).

5.1 Minds and Inquiry

Until now, my aim has been to sketch the overall shape of a fictionalist approach to the mind, rather than focusing on certain aspects of our mental lives in detail. I have also been largely concerned with ordinary talk about the mind, rather than any particular specialist vocabulary. I now wish to focus on some aspects of the mind that are central to inquiry—that is, to the way in which we find out about the world—especially concepts, knowledge, and understanding. Some of the terms that we use to talk about such matters are part and parcel of folk psychology (e.g. 'knowledge', 'understanding'). Others are more commonly used as terms of art in areas such as epistemology or philosophy of science (e.g. 'abstract idea', 'inference'). I want to ask whether mental fictionalism can help us to understand these aspects of our mental life. What are fictionalism's implications for our understanding of the nature of inquiry—and, perhaps, for its limits?

Many of our greatest epistemologists have been committed representationalists. Despite their disagreements, Descartes, Locke, Berkeley, and Hume all accepted the basic vision of the mind as an inner realm of representations. Much of contemporary cognitive science shares this vision—although it adds, of course, its own

Mind as Metaphor: A Defence of Mental Fictionalism. Adam Toon, Oxford University Press. © Adam Toon 2023.
DOI: 10.1093/oso/9780198879626.003.0006

106   MIND AS METAPHOR

technical sophistication and terms of art. Not surprisingly, the representational theory of mind shapes its proponents' understanding of the central issues in epistemology. Each of these issues comes to be seen as a question about the nature of our inner world and its relationship to the world outside. Do the inhabitants of our inner world all stem from experience or are some innate? How do we move from one representation to the next? Is this a rational process? How do those representations that constitute our concepts differ from those that constitute our knowledge or understanding? Finally, and perhaps most alarmingly, how can we know whether our inner representations mirror the world outside the mind—or even whether there is such a world at all?

Representationalists take the folk conception of our inner machinery to be largely accurate. Eliminativists deny this. According to the eliminativist, folk psychology is a false theory, which will eventually be supplanted by developments within cognitive science. Like representationalism, eliminativism also has important implications for epistemology. After all, much of epistemologists' talk about the nature of knowledge and inquiry is built upon the foundations provided by folk psychology. If these foundations are shaky, then the entire edifice threatens to fall. To take an obvious example, according to the traditional analysis, knowledge is justified true belief. If we accept this analysis then it seems that, if beliefs do not exist, neither does knowledge. Accordingly, Paul Churchland (1989) argues that epistemologists and philosophers of science must abandon their existing 'sentential' framework for discussing scientists' concepts, knowledge, and understanding. In its place, they must learn to adopt a radically new framework, in which the central notions are drawn from the technical vocabulary of neural network modelling. According to this approach, for instance, an individual's theory of the world 'is not a large collection or a long list of stored symbolic items. Rather, it is a specific point in that individual's synaptic weight space' (1989, p. 177).

Both representationalists and eliminativists assume that folk psychology, along with its more sophisticated add-ons within epistemology, is an attempt to describe our inner machinery. For his part, Churchland is also content to describe our mental life in terms of inner representations; he simply argues that these representations take a different form to those envisaged by the folk. In Chapter 2, we considered a number of alternative approaches to the mind that each reject the idea of the mind as an inner world, including behaviourism and instrumentalism, along with Ryle and Dennett's views. Unlike eliminativism, each of these approaches would allow us to retain our traditional vocabulary for describing the nature of knowledge and inquiry, even if cognitive science turns out the way that Churchland envisages. And yet their interpretation of this vocabulary will differ sharply from that assumed by the representationalist. Consider the traditional analysis of knowledge once again. Clearly, if knowledge is justified true belief, then different conceptions of the nature of belief will lead to different conception of the nature of knowledge. For a behaviourist, for example, epistemology might be seen as telling us how we ought to act, rather than how to mirror the world.

Especially important for our purposes is Ryle's approach to the intellect, which we encountered briefly in Chapter 1. For the most part, Ryle regards the technical vocabulary of epistemology as having little use in everyday life. Where the epistemologists' terminology does find an echo in our ordinary concepts for describing the intellect, Ryle's analysis typically treats such concepts as dispositional. More specifically, these concepts serve to pick out abilities or capacities of various kinds. For instance, a soldier gradually learning to read contour lines is unlikely to describe himself as acquiring the 'abstract idea of Contour' (Ryle, 1949, 290). We can, if we choose, describe what he is doing in this way. But we should be careful not to fall into the misapprehension that, in doing so, we are describing the slow appearance of some mysterious inner picture. Instead, to say that, at the end of his training, the soldier possesses the abstract idea of Contour would simply be to say that he can now accomplish various tasks. For instance, he can now read maps with contour lines and use them to find his way around the landscape. In a similar vein, Ryle suggests that we understand both knowledge and understanding as kinds of capacities (e.g. 1949, pp. 50–9, 128–30).

We often assume that inquiry should be seen in terms of the activity of the mind. Not all approaches to inquiry see it in this way, however. A famous example is Popper's *Objective Knowledge* (1972). Popper distinguishes between three 'worlds': the 'first world' of material objects, the 'second world' of mental states, and the 'third world' of the '*objective contents of thought*' (1972, 106; emphasis in original). Traditionally, Popper says, epistemologists have focused on the second world. Descartes, Locke, Berkeley, Hume, Kant, and Russell are all 'belief philosophers' (p. 107): each is interested in knowledge in a *subjective* sense, as a state of mind. Popper thinks this is a mistake. In fact, the second world is largely irrelevant for epistemology, at least when it comes to understanding scientific knowledge. Instead, epistemologists ought to focus on the third world. Its inhabitants are not mental states, but theories, problems, arguments, and 'the contents of journals, books, and libraries' (p. 107). It is here that we find knowledge in an *objective* sense or, as Popper puts it, '*knowledge without a knowing subject*' (p. 109; emphasis in original).

Interestingly, Ryle articulates a similar vision for epistemology. When studying the sciences, he argues, epistemologists should confine themselves to 'the systematic study of the structures of built theories' (1949, 299). Such an account—which Ryle suggests we might call the Logic or Grammar of Science—'does not describe or allude to episodes in the lives of individual scientists.... It describes in a special manner what is, or might be, found in print' (1949, 299). According to Ryle, it is here that the epistemologist's favoured terminology (e.g. 'concepts', 'ideas', 'inferring') finds its proper home. The difficulty arises when these terms come to be misapplied to an inner world:

108   MIND AS METAPHOR

the great epistemologists, Locke, Hume, and Kant, were in the main advancing the Grammar of Science, when they thought that they were discussing parts of the occult life-story of persons acquiring knowledge. They were discussing the credentials of sorts of theories, but they were doing this in para-physiological allegories.   (1949, 299)

For Ryle, such 'biographical anecdotes' are simply 'myths' (1949, 267). In what follows, I will argue for a different, but related, approach to epistemology. Like Ryle, I will suggest that many of our key epistemic terms find their home in print—or, more accurately, in the world of public representations of various sorts. And yet, if our application of these terms to an inner world is indeed an allegory, it is an instructive one. For here, as elsewhere, such talk provides us with an invaluable collection of metaphors for describing the activity of the mind. Let us now turn to consider this approach.

5.2 Epistemology as Fiction

How should the fictionalist make sense of inquiry? In Chapter 4, we considered the relationship between minds and material culture. We saw that, for the fictionalist, the two are closely intertwined. We project the world of material culture inwards to make sense of someone's behaviour, and their behaviour can itself rely upon interaction with material culture. Recall Otto and Inga. To say that Inga believes that MoMA is on 53rd Street is to say that she behaves as if she had this written down in a trusty, ever-present—but entirely fictional—notebook inside her head. Otto exhibits the same pattern of behaviour, and thereby possesses the same belief, by interacting with a trusty, ever-present—and entirely real—notebook in his hand. My aim is to offer a similar analysis of concepts, knowledge, and understanding. For each of these aspects of inquiry, I shall argue, we find that our minds and material culture are entangled in a similar manner. To appropriate Popper's terminology for a moment, we might say that our metaphors project the third world of external representations into the second world of the mind. At the same time, states in the second world are often constituted by interactions with the items in the third. This third world is nothing mysterious or ethereal, however: its inhabitants are simply the countless items of material culture that we use to find out about the world, such as field guides, textbooks, or diagrams on a blackboard.

Three points should be noted at the outset. First, in each of the areas I will discuss (i.e. concepts, knowledge, and understanding), we will see that existing approaches can be split into three camps: representationalist views (e.g. Hume's), ability views (e.g. Ryle's), and objective approaches (e.g. Popper's). Of course, this division is somewhat rough and ready, but it is nonetheless helpful. As we will see,

INQUIRING MINDS   109

fictionalism combines aspects of each of these approaches. Second, it is important to acknowledge that the areas I will discuss are each the subject of lengthy and complex debates in their own right. I cannot hope to resolve these debates here. My aim is simply to introduce a fictionalist approach to inquiry and explore some of its implications. Finally, it is important to point out that, in one respect, my discussion of these issues will differ from the approach taken in previous chapters. For most of this book, my aim has been to interpret folk discourse as we find it, not to reform that discourse. The folk, I have argued, are already fictionalists—even if they might not put it quite that way themselves. As we have seen, however, many epistemologists are already committed representationalists. For this reason, much of what I say in this chapter will have to be taken in a revolutionary, rather than hermeneutic, spirit (for this terminology, see Burgess 1983 and Stanley 2001).

5.2.1 Concepts

The literature on concepts is considerable and marked by debates over a range of issues (for an overview, see Margolis and Laurence 1999). Despite these disputes, however, most authors agree that possessing a concept C is typically associated with two different sorts of abilities: *sorting* and *inferring*. Sorting is a matter of being able to distinguish Cs from non-Cs. For example, if Susan has the concept DOG then she can distinguish dogs from cats, tables, chairs, and so on. (For the rest of the discussion, I will follow the standard convention of using capital letters when referring to concepts.) *Inferring* is a matter of being disposed to draw certain inferences from thoughts involving C. In Christopher Peacocke's terms, someone who possesses the concept finds the relevant inferences 'primitively compelling' (1992, 6). For example, if Susan has the concept DOG, then when she thinks that Rover is a dog, she will be disposed to infer that Rover is an animal.

Disagreement begins when we ask: what *are* concepts? Proponents of a *representationalist view* of concepts will argue that concepts are mental representations that underpin these abilities of sorting and inferring (e.g. Margolis and Laurence 2007). For instance, Susan's concept DOG is some sort of mental representation that allows her to distinguish dogs from cats, tables, chairs, and so on, and which leads her to draw certain inferences when she has thoughts about dogs. In contrast, an *ability view* of concepts will eschew talk of mental representations (e.g. Kenny 2010). On this view, to say that Susan possesses the concept DOG is simply to say that she possesses the relevant abilities of sorting and inferring, not to invoke any inner representations that underpin these abilities. Finally, defenders of an *objective view* of concepts will argue that to understand the nature of concepts we should look to representations beyond the mind, especially to language (e.g. Peacocke 1992). In Popper's terms, concepts are inhabitants of the third world, not the second. On this view, the concept DOG is, roughly speaking,

110 MIND AS METAPHOR

the meaning of the word 'dog' in Susan's language. Proponents of such accounts typically view meanings as abstract objects, like Fregean senses—although the question of how we can access these objects remains somewhat mysterious.

How should a fictionalist understand concepts? Let us start by thinking about an external representational device used for sorting and inferring. Consider birdwatching field guides. The use of these guides is the focus of a well-known study by the sociologists Michael Lynch and John Law (1999). Lynch and Law stress that birdwatchers typically go into the field equipped with a range of different tools. For instance, they use powerful binoculars or spotting scopes, and specialized lists to record their sightings and direct their future observations. Especially important, however, are field guides. Lynch and Law focus on Roger Tory Peterson's classic *Field Guide to Western Birds* (1961). The key feature of such guides, of course, is that they are used to identify birds out in the field, rather than read in the comfort of an armchair. As Lynch and Law (1999, 327) put it:

> Peterson's collection is arranged to facilitate the conjoint activity of seeing-and-reading-at-a-glance. The book itself is compact, portable, and durable. Its textual elements are arranged so that a reader can carry it into the field, quickly crack it open after catching a fleeting glimpse of a specimen, and easily derive 'essential' information about the names, descriptions, sizes, and distinctive markings of a set of comparable species.

The guide's illustrations, text, and layout are all carefully designed to allow it to be used in this way. Consider the double-page spread devoted to small wading birds in my copy of Peterson's *A Field Guide to the Birds of Britain and Europe* (1965). On the left-hand side is a colour plate, with closely packed illustrations of sixteen different species of wading birds, such as the Sanderling, Dunlin, and Curlew Sandpiper. As Lynch and Law point out, illustrations in Peterson's guides are schematic, with minimal clutter and background detail. All the birds are shown in the same orientation and small arrows are used to indicate crucial differences between species. For instance, the illustration of the Curlew Sandpiper has two arrows: one indicating its 'decurved' bill (i.e. a bill that curves downwards) and one pointing to its white rump, both features that help to distinguish it from other small waders. Fitting all the illustrations together on one page allows for quick and easy comparison of the different species. The accompanying text is found on the facing page. Once again, the information it provides is minimal and avoids unnecessary detail. The entry for Curlew Sandpiper simply reads '*Autumn*: Grey; decurved bill; white rump' (Peterson et al. 1965, 117). For more detail, the reader is referred to a fuller entry later in the book, but even this remains clipped and concise and is geared towards identification.

What has all this to do with concepts? Put simply, my suggestion is that, just as we treat memory as if it were entries in an inner notebook, so we treat concepts as

INQUIRING MINDS 111

if they were something like entries in an inner field guide. Roughly speaking, to say that someone possesses a concept C is to say that they behave as if they had an inner field guide that enables them to distinguish Cs from non-Cs and make certain inferences about Cs. For instance, to say that Susan possesses the concept DOG is to say that she behaves as if she had an inner field guide that allows her to distinguish dogs from cats, infer that if the object before her is a dog, then it is an animal, and so on. In this way, we can use the metaphor of an inner field guide to pick out an important pattern in Susan's behaviour. This approach has some advantages over existing views of concepts. Unlike representationalism, it avoids worries about the existence of mental representations, since it invokes these entities only as useful fictions. Unlike the abilities view, it explains how the different skills involved in possessing a concept—like sorting and inferring—hang together: despite their differences, each is a skill we would expect from someone using an (actual) field guide. Finally, unlike an objective view, this approach avoids worries about how we can access concepts, while still acknowledging the importance of shared, public representations in concept possession: to possess a concept is to behave as if you were using a shared, public representation—and it is possible for different people to do this to a greater or lesser extent.

The metaphor of an inner field guide picks out a pattern in someone's behaviour. Often, people exhibit this pattern of behaviour without using any external devices. Sometimes, however, they do rely on such devices. In Chapter 4, we saw that Otto and his notebook provide an example of extended belief. In a similar manner, I think, there can be examples of extended concepts. Consider Michaela and Bill. Michaela is an experienced birdwatcher. When she sees a bird at the water's edge, she studies it carefully and decides that it is a Curlew Sandpiper. She also recalls that Curlew Sandpipers are in the family Scolopacidae and breed in Siberia. Pleased at her first sighting of a Curlew Sandpiper that day, she writes it down in her notebook, picks up her binoculars, and begins searching for other species to add to her list. Now consider Bill. Bill is also an experienced birdwatcher. Unlike Michaela, however, Bill is a devotee of Peterson's guides. He never goes out into the field without his trusty Peterson's, along with his binoculars and various lists. When he spots a bird at the water's edge, Bill quickly reaches for his well-thumbed Peterson's guide. Since he's so familiar with his guide and its layout, Bill quickly finds the page for small waders, with its illustrations of different species and their distinctive markings. He finds the bird again in his binoculars and focuses on features picked out by the guide, such as its decurved bill and white rump. After doing so, Bill quickly concludes that the bird is a Curlew Sandpiper. He sees that Curlew Sandpipers are in the family Scolopacidae and that they breed in Siberia. Pleased at his first sighting of a Curlew Sandpiper that day, Bill writes it down in his notebook, picks up his binoculars again and begins looking for other species to add to his list.

112 MIND AS METAPHOR

I think that Bill and his field guide are a good example of extended concepts. Consider the concept CURLEW SANDPIPER. With his field guide in hand, Bill exhibits the behaviour required to possess this concept. He is able to distinguish Curlew Sandpipers from non-Curlew Sandpipers. For example, he can tell a Curlew Sandpiper apart from even highly similar birds, like the Dunlin. Bill can also draw certain inferences regarding Curlew Sandpipers. For example, since Curlew Sandpipers are listed in the section for Scolopacidae and said to breed in Siberia, and since he trusts his Peterson's guide, Bill can infer that, if the bird over there at the water's edge is a Curlew Sandpiper, then it's in the family Scolopacidae and breeds in Siberia. For the fictionalist, the key point about Otto's notebook is that it enables him to exhibit the same overall pattern of behaviour as Inga's on-board biological memory. As a result, Otto may be said to possess the same belief—namely, that MoMA is on 53rd Street. In a similar manner, Bill's field guide enables him to exhibit the same overall pattern of behaviour as Michaela's on-board conceptual resources. It allows him to identify Curlew Sandpipers in the environment and disposes him to make certain inferences about Curlew Sandpipers. As a result, Bill may be said to possess the same concept—namely, the concept CURLEW SANDPIPER. Since Peterson's guide is extensive and Bill is experienced in using it, he probably possesses many other extended concepts, like DUNLIN and COMMON SANDPIPER.

Of course, we will not acquire extended concepts every time we use a field guide, just as don't acquire extended beliefs every time we pick up a notebook. As we saw in Chapter 4, Otto's notebook meets conditions of 'glue and trust' (Clark and Chalmers 1998, 17): whenever it might be relevant, Otto consults his notebook; its contents are readily available to him; and he trusts what it says implicitly. As a result, our metaphor of memory as a trusty, ever-present notebook is especially apt for describing Otto's behaviour. The metaphor will be less apt in other cases—say, if someone uses a notebook once to scribble down the address of MoMA, then leaves it at the bottom of a drawer at home. In this case, we would be unlikely to attribute any extended beliefs to them. Other cases might lie somewhere between the two—say, if someone has MoMA's address stored in their phone, but they often forget to consult it and check their guidebook instead. In this case, we might feel conflicted when asked whether they possess an extended belief. As we saw, the fictionalist need not insist upon a sharp divide here: there might be no obvious point beyond which the metaphor of memory as a notebook ceases to be useful in describing someone's behaviour.

Similar lessons apply in the case of extended concepts. In many ways, Bill's use of his field guide parallels Otto's use of his notebook. When he spots an unusual bird, Bill always consults his Peterson's guide. Because he is so experienced in using the guide, when he sees a bird in his binoculars, Bill can quickly access the relevant pictures and text in his guide and use them to identify the bird in front of him. Bill also places a high degree of trust in his Peterson's guide and does not stop to question it as he makes use of the information it provides. As a result, our metaphor of concepts as a trusty, ever-present field guide is especially apt for describing Bill's behaviour. The metaphor will be less apt in other cases. If I went for a walk and a Curlew Sandpiper flew across my path, my Peterson's guide wouldn't be much use, since it always sits on the shelf in my office, sadly neglected. Even if I did take it with me, it still probably wouldn't be much use, since it is extremely unlikely that I would be able to find the important information and use it to identify the bird before it disappeared out of sight. As Lynch and Law (1999) point out, using a field guide is often a highly frustrating experience for the novice, who finds herself struggling to juggle her binoculars and field guide, find the right page and spot the subtle and elusive marks on the bird in time. Again, we should expect borderline cases: perhaps, after a few months of training using my binoculars and Peterson's guide, I might get to the point where someone would, with some hesitation, be tempted to attribute the concept CURLEW SANDPIPER to me.

A fictionalist approach has implications for other questions that we ask about concepts too. Consider debates over conceptual structure. According to the *classical theory*, concepts are defined by sets of necessary and sufficient conditions. In contrast, the *prototype theory* claims that concepts are applied not via definitions, but by judging similarity to a typical instance of the concept. The *theory theory* rejects each of these views and claims that concepts are like scientific theories. Still other accounts claim that concepts consist of individual *exemplars* of the objects (events, processes, etc.) that they represent or *ideal instances* of those objects (events, processes, etc.). Each of these views excels at explaining some aspects of concept use, while struggling with others. Prototype theories work well for quick, unreflective judgements, like identifying a bird as it flies past us at high speed. They also explain why concepts exhibit *typicality effects*, whereby some instances of the concept are judged to be more central than others (e.g. a robin might be judged a more typical bird than a penguin) (e.g. Rosch and Mervis 1975; Rosch 1978). On the other hand, prototype theories struggle with more reflective judgements, especially about unusual cases (e.g. a white penguin is still a penguin, despite its odd appearance). Such cases are more easily explained if we assume that applying a concept is like invoking a theory (e.g. a theory that says that what makes a bird a penguin is its biological makeup, not the colour of its plumage).

For the representationalist, debates about conceptual structure are debates about the structure of our inner representations: are they definitions, prototypes, theories, exemplars, ideals—or perhaps all of these (Weiskopf 2009)? For the fictionalist, of course, *these* debates will be beside the point, since she denies the existence of inner representations—whether definitions, prototypes, theories, or anything else. And yet the fictionalist can still ask which metaphors best describe our conceptual abilities. Notice that actual, printed field guides—along with other books that fulfil similar purposes, like atlases or catalogues—use lots of different

114 MIND AS METAPHOR

sorts of representations, including many of those invoked in debates over conceptual structure. Some field guides include drawings of a typical instance of a species. Others use photographs of exemplary individuals. Eighteenth-century scientific atlas makers saw it as their job to draw ideal specimens, which might never be found out in the field (Daston and Galison 2007). In some cases, field guides might offer snippets of relevant theories or, more rarely, give definitions of key terms. Such devices provide a rich source of metaphors for describing our conceptual abilities and we need not insist that there is one that is uniquely well suited to do so. It might be that different aspects of concept use are better served by different metaphors.

The situation is rather different if we focus on extended concepts, of course. In these cases, often there *will* be representations that underpin someone's conceptual abilities—although these representations are outside, rather than inside, the head. As a result, it will make sense to ask about the structure of these representations. Our remarks about field guides already suggest that we can expect to find considerable variety here. Consider Bill and Michaela again. Suppose that Michaela's concept SCOLOPACIDAE exhibits certain typicality effects. For example, suppose she is quicker to judge a Curlew Sandpiper as falling into this family than a Temminck's Stint. In this case, it may be that Michaela's concept SCOLOPACIDAE is best described as having a prototype structure. If Peterson's guide simply includes the Curlew Sandpiper and Temminck's Stint on the same page, without making either more prominent or easy to identify, Bill might not exhibit the same typicality effects. His concept SCOLOPACIDAE might more aptly be described as being structured around a set of exemplars. If, rather than his Peterson's guide, Bill were using an eighteenth-century atlas to identify birds instead, his concept might be structured around an ideal specimen. In other cases, different kinds of external devices might result in extended concepts with a classical or theory-like structure. Consider a lawyer who consults reference texts defining obscure legal terms as she decides whether her client's actions constituted fraud. Or imagine a scientist who scribbles down various laws and equations as she determines how to classify a new star.

Fictionalism therefore suggests a kind of *conceptual pluralism*, albeit rather different from that envisaged by some representationalists (Margolis and Laurence 1999; Weiskopf 2009). For the fictionalist, conceptual pluralism simply means: first, that different aspects of concept use might be described by invoking different (metaphorical) representational structures; and second, that different external devices might result in extended concepts with different (actual) representational structures. Fictionalism about concepts has further important implications too, besides this sort of conceptual pluralism. In particular, it suggests that new tools allow us to think new thoughts. I will return to this idea in Section 5.3. For now, let us turn to consider a fictionalist approach to knowledge.

5.2.2 Knowledge

Knowledge—so the traditional analysis goes—is justified true belief. For the *representationalist*, beliefs are inner representations. The traditional analysis therefore suggests that we see knowledge as subset of those representations—namely, those representations that mirror the world correctly and can be backed up by reasons. Ryle offers a rather different *ability view* of knowledge (1949, 128–30). Knowledge, Ryle says, is an ability, or capacity—such as the capacity to answer certain questions correctly, to perform certain tasks appropriately, and so on. In contrast, belief is a tendency to say certain things, act in certain ways, and so on. Unlike the traditional view, Ryle's account allows for the possibility of knowledge without belief (Myers-Schulz and Schwitzgebel 2013). For instance, we might say that a widow *knows* that her husband has died but cannot bring herself to *believe* it: if forced to confront the situation, she can tell you that her husband is dead; and yet in most respects she tends to behave as if he were still alive. In Ryle's terms, she has the relevant capacity, but lacks the associated tendency. Finally, as we saw in Section 5.1, Popper's *objective view* of knowledge identifies it not with the mental state of any individual person—whether understood as an inner representation or a capacity—but with the contents of journals, books, and libraries.

How should the fictionalist understand knowledge? Once again, I suggest that we begin by considering certain sorts of external, public representations, like textbooks, journals, libraries, databases, and the rest. It is these items that provide the inhabitants of Popper's third world, of course. For the moment, let us focus on their role in shared, public practices for creating, storing, and disseminating information. Consider a database. Not just any old information makes it into a database—at least, not if the database is any good. Instead, the information is carefully controlled: it should be suitably justified (or perhaps come from a reliable source) and its contents checked to make sure they are accurate. This information should also be made easy to access, by setting out its entries in a common format, placing them in alphabetical order, using a search function, and so on. If all this is done well, a database enables its users to *do* various things. For instance, with access to their company's database, a member of the sales team at Blogg's Building Supplies might be able to provide you with the details of a particular client—say, Smith's Building Ltd: she could tell you Smith's address, when they last placed an order, how much it cost, and so on. She might also be able to call Smith's, send them a copy of the company's latest catalogue, or try to find out why they haven't placed an order recently. Having access to a database gives its users capacities like these—provided, of course, that they are able to use the database properly and understand what it says.

Earlier I suggested that our talk about concepts is guided by the metaphor of an inner field guide. In similar vein, I suggest that our talk about knowledge is guided

116 MIND AS METAPHOR

by the metaphor of something akin to an inner reference book or database that holds the relevant information. The metaphor of an inner field guide helps us to describe a certain set of human capacities, such as the capacity to distinguish different objects (e.g. tell the difference between a Curlew Sandpiper and a Dunlin) and make inferences about them (e.g. infer that they are in the family Scolopacidae). The metaphor relies upon the fact that these are exactly the sorts of capacities that a field guide gives its readers. Likewise, the metaphor of an inner database helps us to describe another important, but rather different, set of human capacities, such as the capacity to answer questions correctly (e.g. 'What is the address for Smith's Building Ltd?') or perform certain tasks successfully (e.g. send Smith's a new catalogue). In this case, the metaphor relies upon the fact that these are exactly the sorts of capacities that a database gives its users. Of course, one reason why databases give their users these capacities has to do with the nature of their contents: the information in the database should be true and suitably justified (or reliably obtained). If this is not the case, then someone who has access to the database is unlikely to possess the relevant capacities: she will answer questions incorrectly or be unable to carry out certain tasks successfully. Similar constraints apply to our metaphorical, inner database: *its* contents should also be true and suitably justified (or reliably obtained).

On this view, to say that someone knows *p* is to say that they behave as if they had *p* stored in an inner database. What does this mean for the relationship between knowledge and belief? In Chapter 4, I suggested that, to say that someone believes *p* is to say that they behave as if they had *p* written down in an inner notebook. I also suggested that the rules that govern our use of this notebook are summed up well in Clark and Chalmers' 'glue and trust' conditions: our inner notebook is always with us, we trust what it says implicitly, and, whenever its contents might be relevant, we rarely act without consulting it. There are important differences between this metaphor of belief as an inner notebook and the metaphor of knowledge as an inner database. In one respect, the database metaphor is more demanding since its contents must be true and justified (or reliable). These constraints do not apply to our inner notebook: although we typically trust what our inner notebook says, it might still be wrong. We can believe falsehoods, but we cannot know them. In another respect, however, our notebook metaphor is more demanding, since it requires that its owner should consult the notebook whenever its contents might be relevant. The same is not true of our inner database: although someone can consult it if they so wish, they might not do so often. This means that fictionalism allows for the possibility of knowledge without belief: someone can have *p* lodged somewhere in their inner database but fail to consult it regularly enough for it to count among their beliefs. (For a similar point, see Myers-Schulz and Schwitzgebel 2013, 381.)

How does fictionalism relate to existing views of knowledge? Like representationalism, fictionalism allows that the notion of a particular kind of inner representation—namely, one that is true and suitably justified (or reliably obtained)—lies at the heart of our conception of knowledge. But, of course, fictionalism insists that these representations are not real, but metaphorical. And they are not necessarily *beliefs*: as we have seen, fictionalism allows that someone can know *p* without believing *p*. In this respect, fictionalism echoes Ryle's approach: in Ryle's terms, we might say that the metaphor of an inner database is used to identify a capacity, while that of an inner notebook identifies a tendency. At the same time, fictionalism explains why these capacities and tendencies 'operate...in the same field', as Ryle puts it, so that the behaviours they involve 'hang together on a common propositional hook' (1949, 129). For the metaphors that we use to pick out these capacities and tendencies *both* involve the same proposition, whether in a database or notebook. Finally, like Popper's objective view, fictionalism gives a central role to external, public sources of knowledge. What fictionalism adds is the idea that this objective sense of know-ledge shapes our subjective sense of knowledge too.

Can there be cases of extended knowledge? I think there can. For the fiction-alist, these will be cases in which someone displays the capacities required for attributing knowledge by relying, at least in part, upon interacting with an external device. Once we reflect for a moment, we can see that there are lots of cases like this. Consider the database again. Suppose that the company director wants to know the value of Smith Building Ltd's latest order. She tells her secretary, 'Call Tom in the sales team, he'll know.' When she says this, the director is unlikely to assume that Tom knows the figure off by heart. Instead, she expects him to have to look it up in the company's database. Or consider another example. Suppose that you need a friend's address to send them a birthday card. 'Don't worry', I say, 'I know it', and immediately take out my diary to read out the address. You might also ask me if I know their phone number. Again, I reply that I do, but this time I take out my phone to look it up for you. (See Farkas 2015 for a discussion of such cases.) We often attribute knowledge in contexts like these. According to the fictionalist, this is because what we care about is whether the person possesses the relevant capacity, not whether they have to rely on external devices in order to display it.

In Chapter 4, we saw that cases of extended belief can take many different forms. The same is true of extended knowledge. For instance, the external device need not be a linguistic one. I might know where a friend lives, but only if I have a street map with their house marked on it. A detective might know that the suspect has a tattoo because he has his Identikit photo on file. In fact, the external device need not store the relevant information at all. Instead, it might help someone to display the required capacity through other means. For instance, I might know where my friend lives, but only if I have a street map to hand—even if their house is not, in fact, marked on the map. Instead, it could be that simply seeing a map is enough to prompt me to identify their street correctly. Or imagine a scientist who

118 MIND AS METAPHOR

can tell you that a substance has a particular chemical structure, but only if she can first scribble down its formula on a piece of paper or build a ball-and-stick model of it. In other cases, the external device might not even be representational. Once I heard an odd tip for exam revision: revise in a room with a distinctive scent—say, lavender—then make sure you can smell the same scent in the exam hall, and it'll help you remember what you need to know. I've never tried this technique myself but let us suppose that it can make all the difference between being able to recall, say, the year of the Battle of Agincourt in your history exam. If that's right, we might have knowledge that is extended by the scent of lavender.

In some respects, extended knowledge might even be more common than extended belief. (Once again, see Farkas 2015 who makes a similar claim, albeit on different grounds.) What makes Otto so unusual is that he carries his notebook with him wherever he goes and rarely acts without consulting it. Few of us rely on external devices to quite this extent. As we have seen, the conditions for knowledge are less demanding in this respect: someone must be able to produce the information when necessary, but need not do so all the time. This is especially important when it comes to cases of extended knowledge. There are lots of cases in which we rely on external devices to give us the capacity to produce certain information, but where we do not incorporate these devices into our lives as thoroughly as Otto does his notebook. Consider the company database once again. The database might give Tom in the sales team the capacity to produce all sorts of information about the company's clients and sales. In this respect, it might count as a case of extended knowledge. At the same time, it might not be a case of extended belief, since Tom does not have access to the database at all times and, even in cases in which its information is relevant, he often acts without consulting it. For instance, suppose Smith's Building Ltd were based in Nottingham for years, but have recently moved to Derby. The database has been updated accordingly, so Tom can give the correct address if needs be. Still, he can't seem to get it out of his head that they are based in Nottingham: if he's in a meeting and someone mentions Smith's, he says they're in Nottingham; if he needs to visit their office, he still finds himself heading off towards Nottingham; and so on. Here we have a case of extended knowledge, without the corresponding belief. In Ryle's terms, Tom has the required capacity, but not the accompanying tendency.

Of course, not just any old use of an external device will necessarily lead to a case of extended knowledge, just as it won't necessarily lead to an extended belief or an extended concept. If Tom were unable to use the company's database or find the record of Smith's address, then we would be unlikely to say that he knows it. Context is especially important in the case of extended knowledge too. There are particular contexts in which we are interested in someone's capabilities without access to any external devices. Exams and tests are an obvious example (see also Houghton 1997, 168). Consider the famous case of 'The Knowledge'—the daunting amount of information that London taxi drivers are expected to learn concerning approximately 25,000 streets located within six miles of Charing Cross. Drivers are expected to learn this information off by heart and answer questions in their oral exams (e.g. 'Where is the Cheshire Cheese pub?') without recourse to a map or any other external aids. In this particular context, then, attributions of knowledge explicitly rule out dependence upon external devices (although the scent of lavender might be allowed, I suppose). In most other cases, however, our requirements are rather less severe. If you ask a friend if she knows where to find the Cheshire Cheese and she has to glance at a map before pointing it out to you, you would hardly be likely to insist that she didn't know its location after all. Not even exams or tests bar access to external devices in every instance. School pupils are often allowed to take calculators or set texts into their exams with them.

5.2.3 Understanding

Finally, let us turn to consider understanding. Just as in the case of concepts and knowledge, existing accounts of understanding can be divided into representationalist views, ability views, and objective approaches. Most authors agree that, in order to understand some phenomenon, it is not enough merely to believe, or even know, various facts about it. After all, someone can simply learn a list of facts by rote without gaining any understanding. Instead, understanding is often said to require that we 'see' or 'grasp' how these facts fit together. For instance, Catherine Elgin writes that 'to understand the Comanches' dominance of the southern plains involves more than knowing the various truths that belong to a comprehensive, coherent account of the matter. The understander must also grasp how the various truths relate to each other' (Elgin 2009, 323; see also Kvanvig 2003 and Riggs 2003). For our purposes, it is interesting to note that this talk of 'seeing' or 'grasping' remains largely metaphorical (Grimm 2010). A *representationalist view* of understanding must, it seems, move beyond these metaphors to give a proper analysis of this relationship between our inner representations. In contrast, *ability views* aim to identify the distinctive abilities that are characteristic of understanding (e.g. Chang 2009; Ylikoski 2009). For instance, perhaps someone who knows about a topic can recite facts about it, while someone who understands the topic can answer questions about what might have happened if things had been different. Finally, *objective approaches* show little concern with understanding as a psychological phenomenon. Instead, their focus is upon spelling out what makes a good *explanation*, where this is typically seen as a relationship between propositions (e.g. Hempel 1965).

What does a fictionalist approach to understanding look like? As always, we can begin by considering the use of external representations. Earlier I suggested that

120 MIND AS METAPHOR

talk about knowledge relies on the idea of an inner reference book or database that we use to store information. Talk of 'seeing' and 'grasping' how various items of information fit together points towards a rather different, more active use of external representations. Consider spider diagrams (or 'mind maps'). These are diagrams that are used to organize different items of information and show the connections between them. The main topic is presented in the middle of a web, with arrows pointing out towards different aspects of it. Each of these aspects might sit at the centre of its own, smaller web, indicating further related facts or ideas, and so on. Imagine a student taking notes as she reads about the history of the southern plains. The student might simply write a long list of the various items of information that she encounters: that the Comanche were one of the first tribes to acquire horses from the Spanish; that they led a nomadic lifestyle; that they fought on horseback; that they gradually displaced the Apache, and so on. Alternatively, she might try to organize these facts into a spider diagram as reads: she might group them into different topics, draw arrows between them to show their connections, underline them if they seem especially important, and so on. In doing so, the student hopes not merely to record a list of isolated facts about the Comanche, but also to show the relationship between them.

The aim of external representations like spider diagrams is to help us to see complex sets of information at a glance. It is these sorts of representations, I suggest, that lie at the heart of talk about understanding as a matter of 'seeing' or 'grasping' connections between different items of information. To say that someone understands a phenomenon is to say that they behave as if they had something akin to an inner spider diagram showing important facts about it and the relationship between them. Like reference books or databases, spider diagrams enable us to do certain things, but exactly *what* they enable us to do is rather different. For instance, someone who has a database containing a long list of facts about the Comanche can answer questions like 'did the Comanches have horses?' or 'what happened to the Apaches?' In contrast, someone who has a diagram that organizes these facts and illustrates the connections between them enjoys a different, somewhat richer, set of abilities. For instance, they can answer questions like 'why did the Comanches have a nomadic lifestyle?', 'why did the Comanches dominate the southern plains, rather than the Apaches?' or 'how might the history of the plains have been different if the Spanish had never arrived in America?' For the fictionalist, understanding differs from knowledge in roughly the same sorts of ways in which spider diagrams differ from long lists of facts in reference books or databases.

In this way, a fictionalist approach can acknowledge the important idea that understanding involves 'seeing' or 'grasping' the relationship between various facts about its subject matter. At the same time, unlike representationalism, fictionalism need not try to offer an analysis of this seemingly mysterious psychological act. Instead, like much of our talk about the mind, talk of

INQUIRING MINDS 121

understanding in terms of 'seeing' or 'grasping' remains metaphorical—but it is a metaphor that helps to identify certain human abilities. Unlike a straightforward ability view, however, fictionalism can also explain *why* we pick out certain abilities as important for understanding: they are precisely the abilities that follow from possessing a clear overview of the facts in a particular domain and the connections between them. By recognizing this propositional element within our notion of understanding, fictionalism also allows us to ask questions that are difficult to address within an ability view. For instance, we can ask whether the propositions that figure in our understanding of a domain must be true (e.g. Elgin 2009). Finally, fictionalism can share the objective approach's interest in explanation as a public practice, without overlooking understanding as a psychological phenomenon.

Like concepts and knowledge, understanding can also be extended by material devices. In most contexts, when we attribute understanding to someone, what we care about is their overall pattern of behaviour: they must behave as if they have access to the relevant facts and can 'see' the connections between them. Typically, we do not care if displaying this pattern of behaviour relies upon interaction with external devices. In some cases, the role of the external device might be to record items of information. For instance, suppose that a historian has written a long and detailed history of the Comanche and the southern plains. She might be unable to memorize each and every one of the facts that appear in her book—say, the estimated populations of each of the different Comanche bands throughout the course of the eighteenth century. This would hardly lead us to deny that she understood the Comanches' dominance of the southern plains. Indeed, she might be the world's leading authority on the subject. In a similar vein, a chemistry exam might remind students of the structures of certain molecules, since it is interested in testing the students' understanding of their properties, rather than their ability to memorize long lists of chemical formulas.

In other cases, the role of an external device might be not to store the various items of information, but to help someone to 'see' their connections. For instance, consider Dutch roll (Toon 2015). Dutch roll is an oscillatory motion that can affect planes when they fly through turbulence. It is a fairly complicated phenomenon: merely writing down the equations of motion required to explain it takes a page or two, and textbooks normally include a series of diagrams and graphs to show the sequence of steps in a typical Dutch roll cycle. Now suppose that Barbara is an aeronautical engineer. When she is asked why planes experience Dutch Roll, Barbara is able to write down the relevant theoretical principles and facts about air pressure, plane's wings, and so on. She can also show how these principles and facts combine to lead to Dutch roll. To do so, however, Barbara has to resort to pen and paper: she writes down various equations, manipulates them step-by-step, scribbles down a quick sketch of a plane's wing, and so on. Once again, I think we would be happy to attribute understanding in a case like this.

122 MIND AS METAPHOR

Barbara has access to the relevant facts and principles and can 'see' or 'grasp' the connections between them. It is simply that, in her case, achieving this feat involves literal, rather than merely metaphorical, acts of seeing and grasping: she must create and manipulate a collection of external, material representations. Given the complexity of the phenomena that scientists seek to explain, cases like Barbara's might even be the norm, rather than the exception.

5.3 Reason, Tools, and History

An important consequence of a fictionalist approach to inquiry is that it suggests that the mind has a history. Recall a famous passage from the introduction to Hume's *Treatise of Human Nature*:

> 'Tis evident, that all the sciences have a relation, greater or less, to human nature; and that however wide any of them may seem to run from it, they still return back by one passage or another. Even *Mathematics, Natural Philosophy*, and *Natural Religion*, are in some measure dependent on the science of Man; since they lie under the cognizance of men, and are judg'd of by their powers and faculties. 'Tis impossible to tell what changes and improvements we might make in these sciences were we thoroughly acquainted with the extent and force of human understanding, and cou'd explain the nature of the ideas we employ, and of the operations we perform in our reasonings.   (2000, 4)

For Hume, accomplishing this task means discovering the laws that govern the operations of our inner world, much as Newton had discovered those that govern the outer world. For the fictionalist, however, one of the most remarkable features of human inquiry is the creation of new tools and practices that transform the 'extent and force of human understanding' itself.

Consider the sciences. Often, the history of the sciences is told as a succession of theories. And yet many of its key developments turn upon the creation of new tools and new social practices for using those tools. Greek mathematicians created new kinds of diagrams and a new mathematical language with which to construct their proofs (Netz 1999). The members of the early Royal Society created new instruments, new social and literary practices, and new institutions for witnessing their operations (Shapin and Schaffer 1985). At around the same time, mathematicians began to develop new tools for reasoning about probabilities and statistics (Hacking 1990). Eighteenth-century naturalists created new diagrams, tables, and naming systems for classifying plants and animals (Müller-Wille and Charmantier 2012). Nineteenth-century chemists devised new formula systems and new three-dimensional physical models to classify substances and understand their properties (Klein 2003; Rocke 2010). Twentieth-century physicists built new

INQUIRING MINDS 123

instruments and devised computer simulations to investigate the subatomic world, along with new institutions to make sense of the images and data they produced (Galison 1997).

On the face of it, Cartesianism struggles to do justice to these extraordinary innovations in the nature of human inquiry. Put simply, if scientists' concepts, knowledge, and understanding are found inside their heads, then the material and social dimensions of inquiry must fall outside the domain of thought and reason. At best, new tools and practices might allow scientists to express their thoughts or help them in their reasoning, but they are not part of the operations of the mind. Popper's objective approach fares little better. Popper is careful to point out that the third world is not merely an expression of scientists' inner states: although a human construct, it exists largely independently from the second world of the mind and has its own dynamics (1972, 147). And yet Popper's conception of the third world as the '*objective contents of thought*' (1972, 106; emphasis in original) does little to place it within its material and social context (Bloor 1974). Like the Cartesian, Popper also retains a sharp dividing line between mind and material culture: although the second and third worlds might interact, they remain distinct (cf. Houghton 1997, 166–7).

In contrast, a fictionalist approach to inquiry shows that new tools and social practices can fundamentally alter the nature of human thought and reason. This can happen in three main ways.

First, new tools can enable new patterns of behaviour. Most obvious here are linguistic tools. As we saw in Chapter 3, fictionalism need not deny the possibility of thought without language. A baby can want milk even if it cannot say so. And yet, in the case of more sophisticated thoughts, often we cannot display the required pattern of behaviour without having the right words to use. Someone cannot believe that electrons are negatively charged (or that there are an infinite number of primes, or that gross national product is increasing) without possessing the right vocabulary from physics (or mathematics, or economics). Non-linguistic tools can have a similar effect. A scientist who understands the properties of electrons is not only able to talk about them; she also sets up her cathode ray tube correctly, directs its beam using a deflection coil, and so on. And yet, just as there are things you cannot say without the right words, so there are things you cannot do without the right tools. You cannot set up a cathode ray tube correctly if there are no cathode ray tubes. (For an important, and related, set of arguments that focus on the way in which new reasoning practices render new claims meaningful, see Hacking 1982, 1992.)

Second, new tools and practices bring new metaphors for describing behaviour. I have focused on a few central metaphors: notebooks for memory, reference books or databases for knowledge, and so on. The full story is much more complicated, of course. On the one hand, fictionalism suggests that societies without these tools, such as preliterate societies, will have a very different concept

124   MIND AS METAPHOR

of mind from our own. On the other hand, all sorts of other inventions have also been used as metaphors for the operations of the mind (or brain). Searle (1984, 44) offers the following list: telephone switchboards, telegraph systems, hydraulic systems, electro-magnetic systems, mills, catapults, and, most prominent today, the computer. (For a historical account that focuses on memory, see Draaisma 2000.) Not all these metaphors are useful, of course. Sometimes, they prove to be misleading. For instance, Ryle objects that our talk about the intellect is distorted by a contemplative idiom based on the experience of learning geometry by following proofs on a blackboard (1949, 288). For the fictionalist, the key test for any such metaphor is whether it captures patterns in human behaviour. If it proves its worth and becomes entrenched in our language, a new metaphor will reshape our conception of the mind and, in turn, can alter the way that we behave as a result (see also Hacking 1986).

Finally, new tools and practices can enable new extended mental states. In doing so, they allow us to think new thoughts (Houghton 1997; Clark 2006; Kirsh 2010). We must be careful here, of course. In some cases, tools simply enable one person to do what another can do perfectly well without them. I cannot move heavy rocks without a crowbar, but some people can. In other cases, tools exceed the capacities of any individual human body. I doubt whether anyone can drive a screw into metal with their bare hands, however strong they are. The same lesson holds for tools of inquiry. Often, these tools simply enable an individual to do what others can manage on their own. I cannot multiply three-digit numbers without pen and paper, but some people can. Otto cannot remember addresses without his notebook, but Inga can. In other cases, it seems that interacting with external tools and devices allows us to acquire capacities that lie beyond the reach of our bare brains and bodies. Of course, determining the limits of human capacities—for performing mental arithmetic, say, or memorizing addresses—is an empirical matter and cannot be done from the armchair (Kirsh 2010). Despite this, I think that we can still see ways in which new tools might bring new concepts, knowledge, and understanding.

Consider concepts. Bill cannot identify Curlew Sandpipers without his field guide, but Michaela can. This means that the concept CURLEW SANDPIPER is extended for some people, but not for others. In other cases, however, it seems unlikely that anyone can perform the required acts of classification and inference without relying on external devices. Take beetles, for instance. Over 400,000 species of beetles have been discovered. It may be that even experts cannot distinguish some of these species without relying on tools such as field guides and computer databases. If that's right, then the concepts that identify those species are always extended: without the right tools, we cannot possess them. Or consider knowledge. Tom needed to rely on his company's database to know its clients' details. Perhaps one of Tom's colleagues can memorize this information instead. And yet presumably there are some databases that surpass human capacities for memory. Of course, it might be argued that, even if not all a database's information can be known without it, each individual entry can be. But think about a database of protein structures. Macromolecules like proteins can consist of thousands of atoms arranged in a huge variety of extraordinarily complex structures. Knowing the structure of a protein requires being able to identify its amino acids, specify what sequence they appear in, indicate how they are folded, and so on. For some proteins, it may be that no scientist can do these things without access to a database, which gives the protein's chemical formula or displays an image of its structure. The same points hold for understanding. Unlike Barbara, there might be an aeronautical engineer who can 'see' how various facts and principles combine to produce Dutch roll without resorting to pen and paper. But some chains of calculation will be too intricate even for the most accomplished mathematicians. Indeed, many equations cannot be solved analytically, meaning that scientists must resort to computer simulations to explore their implications.

The upshot is that fictionalism differs markedly from the representationalist approach that has underpinned so much of our thinking about knowledge and inquiry. Instead of trying to identify the laws that govern our inner world, we should focus on our interactions with the outer world of material culture. The nature of these interactions, and the metaphors that we use to describe them, are not fixed—both will change as we develop new tools and new social practices to use them. As a result, fictionalism will take a different approach to many questions that have long occupied epistemologists. Consider debates over empiricism. Recall Hume's famous copy principle, which claims that each of our simple ideas is copied from a simple impression. If the copy principle is understood as a claim about inner pictures, the fictionalist cannot accept it. But we might try to reformulate the copy principle along fictionalist lines. According to the fictionalist, to say that someone possesses a concept C is to say that they behave as if they had an inner field guide that enables them to distinguish Cs from non-Cs and make certain inferences about Cs. A revised copy principle might claim that our conceptual capacities, if understood in this manner, depend upon our previous experiences (cf. Ryle 1949, 256–7). For instance, it might claim that we cannot distinguish between things that are blue and those that are not unless we have already seen something blue before.

Is this revised copy principle true? This is a question about the nature of human capacities and their relation to our previous experience. If we want to answer such questions, fictionalism reminds us to reflect upon the way in which human capacities can be transformed by the tools that we use. Think of paint samples. I have a paint sample card in front of me that includes over twenty different shades of blue paint. I doubt whether many people could tell the difference between, say, *Mineral Mist*, *Misty Sky*, and *Frosted Lake* unless they had the sample card in front of them. With the sample card in hand, however, they

126 MIND AS METAPHOR

might be able to distinguish these colours, by holding the card up against the wall and judging which colour matches best. Here we have an example of extended concepts: by using the sample card, we are able to perform acts of classification and inference that were previously beyond our grasp. Of course, this example is no threat to the copy principle. After all, if we have the sample card in our hand, we can see the colours on it. But now consider Hume's infamous 'missing shade of blue'. Hume asks us to imagine

> a person to have enjoy'd his sight for thirty years, and to have become perfectly well acquainted with colours of all kinds, excepting one particular shade of blue, for instance, which it never has been his fortune to meet with. Let all the different shades of that colour, except that single one, be plac'd before him, descending gradually from the deepest to the lightest; 'tis plain, that he will perceive a blank, where that shade is wanting, and will be sensible, that there is a greater distance in that place betwixt the contiguous colours, than in any other. Now I ask, whether 'tis possible for him, from his own imagination, to supply this deficiency, and raise up to himself the idea of that particular shade, tho' it had never been convey'd to him by his senses? I believe there are few but will be of opinion that he can; and this may serve as a proof, that the simple ideas are not always deriv'd from the correspondent impressions.   (2000, 10)

For Hume, this exception to the copy principle is 'so particular and singular, that 'tis scarce worth our observing' (2000, 10). For the fictionalist, however, Hume's missing shade of blue is an instance of a much wider phenomenon—namely, the way in which we can expand our conceptual capacities by using external tools. In effect, the person in Hume's thought experiment is given a paint sample card. In one respect, it is an unusual sample card, since it has a colour missing. And yet, despite this apparent shortcoming, the paint card still expands his conceptual capacities, including those capacities that relate to the missing shade. This is certainly remarkable, but it is not so unusual. In specialist paint shops, customers can ask the supplier to mix bespoke colours that lie somewhere between those shown on the card (e.g. 'I'd like something darker than *Mineral Mist*, but not as dark as *Frosted Lake*'). It may well be the case that, without using the sample card, the customer could not conceive of the colour they wanted. With the card to hand, however, the task becomes relatively easy. In this way, the use of tools might even expand our capacities beyond the constraints envisaged by empiricism.

5.4 Conclusion

Our vision of inquiry is haunted by the idea of the mind as an inner world forever trying (and, perhaps, failing) to reach the world outside. Fictionalism suggests a different vision of inquiry. The mind is not a timeless inner world. It is a metaphorical projection of an outer world of human tools and practices that is constantly changing. These new tools and practices can bring new patterns of behaviour, new metaphors to identify those patterns, and new extended mental states. To use Hume's words, the 'mental geography' of the mind is not fixed but expanding. By creating new tools and practices of inquiry, we have vastly increased the domain of things that lie 'under the cognizance of men'.

Epilogue

An Uneasy Presence...

Some ghosts are thoroughly malevolent. They do nothing but cause trouble and it would be better for everyone if they could be laid to rest. Others are less straightforward characters. They can be unnerving, certainly. And yet somehow we feel that they are with us because they still have their own mysterious purpose to fulfil. The ghost in the machine is like that. It can undoubtedly lead us astray if we begin to take it too seriously. And yet we cannot stop telling our stories about it, for it provides our best way of making sense of ourselves and our place in the world.

Long ago, many philosophers accepted the ghost on its own terms. Its presence was to be granted as something otherworldly, or at least not of the material world. The ghost's remarkable abilities—to describe or picture the world outside, or perform feats of telekinesis to animate the machine it haunted—might be accepted as part and parcel of its ethereal character. Those of a more nervous disposition attempted to exorcize the ghost entirely. But their attempts to make do without its presence were ultimately doomed to failure.

Nowadays, most philosophers try not to exorcize the ghost, but to tame it. The ghost is to be made flesh and become a respectable citizen of our scientific age. Our attitude is rather like that of the no-nonsense father in Oscar Wilde's *The Canterville Ghost*, who offers Sir Simon de Canterville's spirit Tammany Rising Sun Lubricator to oil his ghostly chains. Like Sir Simon, however, the ghost in the machine has resisted our attempts to domesticate it. Despite our best efforts, its character remains mysterious and paradoxical.

Most ghost stories are best left as just that—*stories*. I suggest we do the same with the story of the ghost in the machine. The story can mislead us. Indeed, it *has* misled us and no doubt it will continue to do so. If we stop to interrogate it too closely, we find that it is full of inconsistencies and awkward plot holes. And yet despite all this, we tell the story for a reason. Like many tall tales, the story of the ghost in the machine is part of the way in which we understand the world and find our way around it.

That we can tell stories at all depends upon our nature as social creatures. And it is our social, norm-governed use of tools—especially language—that provides

Mind as Metaphor: A Defence of Mental Fictionalism. Adam Toon, Oxford University Press. © Adam Toon 2023.
DOI: 10.1093/oso/9780198879626.003.0007

the story of the ghost in the machine with its main characters and plot. This plot is not fixed like the text of a novel. Instead, it is more like a folk story, which is continually altered and embellished as it passes from person to person. By now, we have told the story so often that it has become part of who we are. No wonder we tend to forget that, when all is said and done, it is still only a story.

References

Adams, F. & Aizawa, K. (2001). The Bounds of Cognition. *Philosophical Psychology*, 14, 43–64.

Adams, F. & Aizawa, K. (2008). *The Bounds of Cognition*. Oxford: Blackwell.

Appiah, K. A. (2017). *As If: Idealization and Ideals*. Cambridge, MA: Harvard University Press.

Armour-Garb, B. & Woodbridge, J. A. (2015). *Pretense and Pathology*. Cambridge: Cambridge University Press.

Baker, L. R. (1987). *Saving Belief: A Critique of Physicalism*. Princeton, NJ: Princeton University Press.

Baker, L. R. (1994). Attitudes as Nonentities. *Philosophical Studies*, 76(2–3), 175–203.

Beardsley, M. (1962). The Metaphorical Twist. *Philosophy and Phenomenological Research*, 22(3), 293–307.

Bennett, M. R. & Hacker, P. M. S. (2003). *Philosophical Foundations of Neuroscience*. Oxford: Blackwell.

Bennett, M. R. & Hacker, P. M. S. (2007). The Conceptual Presuppositions of Cognitive Neuroscience: A Reply to Critics. In M. Bennett, D. Dennett, P. Hacker, & J. Searle (eds), *Neuroscience and Philosophy: Brain, Mind, and Language* (pp. 127–62). New York: Columbia University Press.

Block, N. (1981). Psychologism and Behaviorism. *The Philosophical Review*, 90(1), 5–43.

Bloor, D. (1970). Is the Official Theory of Mind Absurd? *The British Journal for the Philosophy of Science*, 21(2), 167–83.

Bloor, D. (1974). Popper's Mystification of Objective Knowledge. *Science Studies*, 4(1), 65–76.

Bourne, C. & Bourne, E. C. (2020). Folk Stories. In B. Armour-Garb & F. Kroon (eds), *Fictionalism in Philosophy* (pp. 168–86). Oxford: Oxford University Press.

Braddon-Mitchell, D. & Jackson, F. (2007). *Philosophy of Mind and Cognition: An Introduction*. Oxford: Blackwell (2nd edition).

Brandom, R. (1994). *Making it explicit: Reasoning, representing, and discursive commitment*. Cambridge, MA: Harvard University Press.

Brandom, R. (1997). Study Guide. In W. Sellars (ed.), *Empiricism and the Philosophy of Mind*. Cambridge, MA: Harvard University Press.

Burgess, J. P. (1983). Why I Am Not a Nominalist. *Notre Dame Journal of Formal Logic*, 24(1), 93–105.

Byrne, A. (2012). Review of *The Opacity of Mind: An Integrative Theory of Self-Knowledge*, by Peter Carruthers. *Notre Dame Philosophical Reviews*. Retrieved from https://ndpr.nd.edu/reviews/the-opacity-of-mind-an-integrative-theory-of-self-knowledge/.

Carruthers, P. (2009). How We Know Our Own Minds: The Relationship between Mindreading and Metacognition. *Behavioral and Brain Sciences* 32(2), 121–38.

Carruthers, P. (2011). *The Opacity of Mind: An Integrative Theory of Self-Knowledge*. Oxford: Oxford University Press.

Cartwright, N. (1983). *How the Laws of Physics Lie*. Oxford: Clarendon Press.

132 REFERENCES

Chang, H. (2009). Ontological Principles and the Intelligibility of Epistemic Activities. In H. de Regt, S. Leonelli, & K. Eigner (eds), *Scientific Understanding: Philosophical Perspectives* (pp. 64–82). Pittsburgh: Pittsburgh University Press.

Churchland, P. M. (1981). Eliminative Materialism and the Propositional Attitudes. *The Journal of Philosophy*, 78(2), 67–90.

Churchland, P. M. (1989). *A Neurocomputational Perspective: The Nature of Mind and the Structure of Science*. Cambridge, MA: MIT Press.

Churchland, P. S. (1986). *Neurophilosophy: Toward a Unified Science of the Mind/Brain*. Cambridge, MA: MIT Press.

Clark, A. (2006). Material Symbols. *Philosophical Psychology*, 19(3), 291–307.

Clark, A. (2008). *Supersizing the Mind: Embodiment, Action, and Cognitive Extension* Oxford: Oxford University Press.

Clark, A. (2010a). *Memento's* Revenge: The Extended Mind, Extended. In R. Menary (ed.), *The Extended Mind*. Cambridge, MA: MIT Press.

Clark, A. (2010b). Coupling, Constitution, and the Cognitive Kind: A Reply to Adams and Aizawa. In R. Menary (ed.), *The Extended Mind*. Cambridge, MA: MIT Press.

Clark, A. & Chalmers, D. (1998). The Extended Mind. *Analysis*, 58(1), 7–19.

Colombetti, G. & Roberts, T. (2015). Extending the Extended Mind: The Case for Extended Affectivity. *Philosophical Studies*, 172, 1243–63.

Crane, T. (2016). *The Mechanical Mind: A Philosophical Introduction to Minds, Machines and Mental Representation*. 3rd edition. Abingdon, Oxfordshire: Routledge.

Crane, T. & Farkas, K. (2022). Mental Fact and Mental Fiction. In T. Demeter, T. Parent, & A. Toon (eds), *Mental Fictionalism: Philosophical Explorations* (pp. 303–19). Abingdon, Oxfordshire: Routledge.

Crimmins, M. (1998). Hesperus and Phosphorus: Sense, Pretense, and Reference. *Philosophical Review* 107, 1–47.

Cummins, R. (1989). *Meaning and Mental Representation*. Cambridge, MA: MIT Press.

Daston, L. & Galison, P. (2007). *Objectivity*. Brooklyn, NY: Zone Books.

Demeter, T. (2013). Mental Fictionalism: The Very Idea. *The Monist*, 96(4), 483–504.

Dennett, D. (1969). *Content and Consciousness*. London: Routledge & Kegan Paul.

Dennett, D. (1971). Intentional Systems. *The Journal of Philosophy*, 68(4), 87–106.

Dennett, D. (1987). *The Intentional Stance*. Cambridge, MA: MIT Press.

Dennett, D. (1991a). *Consciousness Explained*. Boston: Little, Brown & Company.

Dennett, D. (1991b). Real Patterns. *The Journal of Philosophy*, 88(1), 27–51.

Dennett, D. (1996). *Kinds of Minds: Toward an Understanding of Consciousness*. New York: Basic Books.

Dennett, D. (2007). Philosophy as Naïve Anthropology: Comment on Bennett and Hacker. In M. Bennett, D. Dennett, P. Hacker, & J. Searle (eds), *Neuroscience and Philosophy: Brain, Mind, and Language* (pp. 73–96). New York: Columbia University Press.

Dennett, D. (2009). Intentional systems theory. In B. P. McLaughlin, A. Beckermann, & S. Walter (eds), *The Oxford Handbook of Philosophy of Mind* (pp. 339–50). Oxford: Oxford University Press.

Dennett, D. (2013). *Intuition Pumps and Other Tools for Thinking*. London: Penguin.

Dennett, D. (2022). Am I a Fictionalist? In T. Demeter, T. Parent, & A. Toon (eds), *Mental Fictionalism: Philosophical Explorations* (pp. 352–62). Abingdon, Oxfordshire: Routledge.

Di Paolo, E. (2009). Extended Life. *Topoi*, 28(1), 9.

Draaisma, D. (2000). *Metaphors of Memory: A History of Ideas about the Mind*. Cambridge: Cambridge University Press.

REFERENCES 133

Drayson, Z. (2022). What We Talk About When We Talk About Mental States. In T. Demeter, T. Parent, & A. Toon (eds), *Mental Fictionalism: Philosophical Explorations* (pp. 147–59). Abingdon, Oxfordshire: Routledge.

Dretske, F. (1981). *Knowledge and the Flow of Information.* Cambridge, MA: MIT Press.

Eklund, M. (2019). Fictionalism. In E. N. Zalta (ed.), *The Stanford Encyclopedia of Philosophy* (Winter 2019 edition). Retrieved from http://plato.stanford.edu/archives/win2019/entries/fictionalism

Elgin, C. (2009). Is Understanding Factive? In D. Pritchard, A. Millar, & A. Haddock (eds), *Epistemic Value* (pp. 322–9). Oxford: Oxford University Press.

Farkas, K. (2012). Two Versions of the Extended Mind Thesis. *Philosophia*, 40, 435–47.

Farkas, K. (2015). Belief May Not Be a Necessary Condition for Knowledge. *Erkenntnis*, 80(1), 185–200.

Fine, A. (1993). Fictionalism. *Midwest Studies in Philosophy*, 18(1), 1–18.

Fodor, J. (1974). Special Sciences: Or the Disunity of Science as a Working Hypothesis. *Synthese*, 28, 97–115.

Fodor, J. (1975). *The Language of Thought.* Cambridge, MA: Harvard University Press.

Fodor, J. (1987). *Psychosemantics: The Problem of Meaning in the Philosophy of Mind.* Cambridge, MA: MIT Press.

Fodor, J. (1989). Making Mind Matter More. *Philosophical Topics*, 17(1), 59–79.

Fodor, J. (2009). Where Is My Mind? *London Review of Books*, 31(3), 13–15.

Fodor, J. & Pylyshyn, Z. W. (1988). Connectionism and Cognitive Architecture: A Critical Analysis. *Cognition*, 28(1–2), 3–71.

Friend, S. (2007). Fictional Characters. *Philosophy Compass*, 2(2), 141–56.

Galison, P. (1997). *Image and Logic: A Material Culture of Microphysics.* Chicago: University of Chicago Press.

Gentner, D. & Bowdle, B. (2008). Metaphor as Structure-Mapping. In R. Gibbs (ed.), *The Cambridge Handbook of Metaphor and Thought.* Cambridge: Cambridge University Press.

Giere, R. (1988). *Explaining Science: A Cognitive Approach.* Chicago: University of Chicago Press.

Godfrey-Smith, P. (2005). Folk Psychology as a Model, *Philosopher's Imprint*, 5(6), 1–16.

Grice, H. P. (1989). *Studies in the Way of Words.* Cambridge, MA: Harvard University Press.

Grimm, S. (2010). The Goal of Explanation. *Studies in History and Philosophy of Science.* 41(4): 337–44.

Hacking, I. (1982). Language, Truth and Reason. In M. Hollis and S. Lukes (eds.), *Rationality and Relativism* (pp. 48–66). Cambridge, MA: MIT Press.

Hacking, I., 1986. Making Up People. In T. C. Heller, M. Sosna, & D. E. Wellbery (eds), *Reconstructing Individualism: Autonomy, Individuality, and the Self in Western Thought* (pp. 222–36). Stanford: Stanford University Press.

Hacking, I. (1990). *The Taming of Chance.* Cambridge: Cambridge University Press.

Hacking, I. (1992). 'Style' for Historians and Philosophers. *Studies in History and Philosophy of Science Part A*, 23(1), 1–20.

Hall, S. (2014). Pass the Ammunition: A Short Etymology of 'Blockbuster'. In A. Elliott (ed.), *The Return of the Epic Film* (pp. 147–66). Edinburgh: Edinburgh University Press.

Hampshire, S. (1950). Review of *The Concept of Mind. Mind*, 59 (234), 237–55.

Haugeland, J. (1985). *Artificial Intelligence: The Very Idea.* Cambridge, MA: MIT Press.

Haugeland, J. (1990). The Intentionality All-Stars. *Philosophical Perspectives*, 4, 383–427.

Hempel, C. (1965). *Aspects of Scientific Explanation and Other Essays.* New York: Free Press.

134   REFERENCES

Hills, D. (2017). Metaphor. In E. N. Zalta (ed.), *The Stanford Encyclopedia of Philosophy* (Fall 2017 Edition). Retrieved from http://plato.stanford.edu/archives/fall2017/entries/metaphor.

Houghton, D. (1997). Mental Content and External Representations. *The Philosophical Quarterly*, 47(187), 159–77.

Hume, D. (2000). *A Treatise of Human Nature*. Oxford: Oxford University Press. (Edited by David Fate Norton & Mary J. Norton. Original work published 1739.)

Hutto, D. (2013). Fictionalism about Folk Psychology. *The Monist*, 96(4), 582–604.

Hutto, D. (2022a). A Brickhouse Defence for Folk Psychology: How to Defeat 'Big Bad Wolf' Eliminativism. In T. Demeter, T. Parent, & A. Toon (eds), *Mental Fictionalism: Philosophical Explorations* (pp. 160–83). Abingdon, Oxfordshire: Routledge.

Hutto, D. (2022b). Getting Real about Pretense: A Radical Enactivist Proposal. *Phenomenology and the Cognitive Sciences*, 21, 1157–75.

Hutto, D. & Myin, E. (2013). *Radicalizing Enactivism: Basic Minds without Content*. Cambridge, MA: MIT Press.

Hutto, D. & Myin, E. (2017). *Evolving Enactivism: Basic Minds Meet Content*. Cambridge, MA: MIT Press.

Jaynes, J. (1976). *The Origin of Consciousness in the Breakdown of the Bicameral Mind*. Boston: Houghton Mifflin.

Joyce, R. (2005). Moral Fictionalism. In M. E. Kalderon (ed.), *Fictionalism in Metaphysics* (pp. 287–313). Oxford: Oxford University Press.

Joyce, R. (2013). Psychological Fictionalism, and the Threat of Fictionalist Suicide. *The Monist*, 96(4), 517–38.

Kirsh, D. (2010). Thinking with External Representations. *AI and Society*, 25, 441–54.

Klein, U. (2003). *Experiments, Models, Paper Tools: Cultures of Organic Chemistry in the Nineteenth Century*. Stanford: Stanford University Press.

Kroon, F. (2001). Fictionalism and the Informativeness of Identity. *Philosophical Studies*, 106, 197–225.

Kvanvig, J. (2003). *The Value of Knowledge and the Pursuit of Understanding*. Cambridge: Cambridge University Press.

Lakoff, G. & Johnson, M. (1980). *Metaphors We Live By*. Chicago: University of Chicago Press.

Langland-Hassan, P. (2014). What It Is to Pretend. *Pacific Philosophical Quarterly*, 95(3), 397–420.

Langland-Hassan, P. (2022). Why Pretense Poses a Problem for 4E Cognition (and How to Move Forward). *Phenomenology and the Cognitive Sciences*, 21, 1003–1021.

Levy, A. (2018). Modeling and Realism: Strange Bedfellows? In J. Saatsi (ed.), *The Routledge Handbook of Scientific Realism*. Abingdon, Oxfordshire: Routledge.

Lipton, P. (2004). *Inference to the Best Explanation*. 2nd Edition. Abingdon, Oxfordshire: Routledge.

Lipton, P. (2005). Testing Hypotheses: Prediction and Prejudice. *Science*, 307(5707), 219–21.

Lycan, W. G. (2022). A Rylean Mental Fictionalism. In T. Demeter, T. Parent, & A. Toon (eds), *Mental Fictionalism: Philosophical Explorations* (pp. 70–85). Abingdon, Oxfordshire: Routledge.

Lynch, M. & Law, J. (1999). Pictures, Texts, and Objects: The Literary Language Game of Bird-watching. In M. Biagioli (ed.), *The Science Studies Reader* (pp. 317–41). Abingdon, Oxfordshire: Routledge.

Magidor, O. (2022). Category Mistakes. In Edward N. Zalta & Uri Nodelman (eds), *The Stanford Encyclopedia of Philosophy* (Fall 2022 Edition). Retrieved from https://plato.stanford.edu/archives/fall2022/entries/category-mistakes.

REFERENCES   135

Margolis, E. & Laurence, S. (1999). *Concepts: Core Readings*. Cambridge, MA: MIT Press.
Margolis, E. & Laurence, S. (2007). The Ontology of Concepts-Abstract Objects or Mental Representations? *Noûs*, 41(4), 561–93.
McGeer, V. (2007). The Regulative Dimension of Folk Psychology. In D. Hutto & M. Ratcliffe (eds), *Folk Psychology Re-Assessed*. Dordrecht: Springer.
McGinn, M. (1997). *The Routledge Guidebook to Wittgenstein and the Philosophical Investigations*. Abingdon, Oxfordshire: Routledge.
Melser, D. (2004). *The Act of Thinking*. Cambridge, MA: MIT Press.
Menary, R. (2007). *Cognitive Integration: Mind and Cognition Unbounded*. Basingstoke: Palgrave Macmillan.
Menary, R. (ed.) (2010). *The Extended Mind*. Cambridge, MA: MIT Press.
Millikan, R. (1984). *Language, Thought and other Biological Categories*. Cambridge, MA: MIT Press.
Moran, R. (1989). Seeing and Believing: Metaphor, Image, and Force, *Critical Inquiry*, 16(1), 87–112.
Morgan, M. & Morrison, M. (1999). *Models as Mediators*. Cambridge: Cambridge University Press.
Müller-Wille, S. & Charmantier, I. (2012). Natural History and Information Overload: The Case of Linnaeus. *Studies in History and Philosophy of Science Part C*, 43(1), 4–15.
Myers-Schulz, B. & Schwitzgebel, E. (2013). Knowing that P without Believing that P. *Noûs*, 47(2), 371–84.
Netz, R. (1999). *The Shaping of Deduction in Greek Mathematics*. Cambridge: Cambridge University Press.
Papineau, D. (1987). *Reality and Representation*. Oxford: Blackwell.
Parent, T. (2013). In the Mental Fiction, Mental Fictionalism Is Fictitious. *The Monist*, 96(4), 605–21.
Peacocke, C. (1992). *A Study of Concepts*. Cambridge, MA: MIT Press.
Peterson, R. T., Mountfort, G., & Hollom, P. A. D. (1965). *A Field Guide to the Birds of Britain and Europe*. London: Collins. (First edition published 1954.)
Popper, K. (1972). *Objective Knowledge: An Evolutionary Approach*. Oxford: Oxford University Press.
Putnam, H. (1981). *Reason, Truth and History*. Cambridge: Cambridge University Press.
Ramberg, P. J. (2003). *Chemical Structure, Spatial Arrangement: The Early History of Stereochemistry, 1874-1914*. Aldershot: Ashgate.
Ramsey, W., Stich, S. P., & Garon, J. (1990). Connectionism, Eliminativism and The Future of Folk Psychology. *Philosophical Perspectives*, 4, 499–533.
Reichenbach, H. (1938). *Experience and Prediction*. Chicago: University of Chicago Press.
Riggs, W. (2003). Understanding 'Virtue' and the Virtue of Understanding. In M. DePaul and L. Zagzebski (eds), *Intellectual Virtue: Perspectives from Ethics and Epistemology* (pp. 203–26). Oxford: Oxford University Press.
Robbins, P. & Aydede, M. (2009). *The Cambridge Handbook of Situated Cognition*. Cambridge: Cambridge University Press.
Rocke, A. J. (2010). *Image and Reality: Kekulé, Kopp, and the Scientific Imagination*. Chicago: University of Chicago Press.
Rosch, E. (1978). Principles of Categorization. In E. Rosch & B. Lloyd (eds), *Cognition and Categorization* (pp. 27–48). Hillsdale, NJ: Lawrence Erlbaum Associates.
Rosch, E. & Mervis, C. (1975). Family Resemblances: Studies in the Internal Structure of Categories, *Cognitive Psychology*, 7: 573–605.

136 REFERENCES

Ross, A. (2022). Mental Fictionalism: The Costly Combination of Magic and the Mind. In T. Demeter, T. Parent, & A. Toon (eds), *Mental Fictionalism: Philosophical Explorations* (pp. 184–98). Abingdon, Oxfordshire: Routledge.

Rowlands, M. (1999). *The Body in Mind: Understanding Cognitive Processes*. Cambridge: Cambridge University Press.

Rupert, R. (2004). Challenges to the Hypothesis of Extended Cognition. *The Journal of Philosophy*, 101(8), 389–428.

Ryle, G. (1949). *The Concept of Mind*. London: Hutchinson.

Ryle, G. (1979). *On Thinking*. (Edited by K. Kolenda.) New Jersey: Rowman & Littlefield.

Schwitzgebel, E. (2001). In-between Believing. *The Philosophical Quarterly*, 51(202), 76–82.

Schwitzgebel, E. (2002). A Phenomenal, Dispositional Account of Belief. *Noûs*, 36(2), 249–75.

Searle, J. (1984). *Minds, Brains and Science*. Cambridge, MA: Harvard University Press.

Sellars, W. (1956). Empiricism and the Philosophy of Mind. In H. Feigl & M. Scriven (eds.) *The Foundations of Science and the Concepts of Psychology and Psychoanalysis*. Minnesota Studies in the Philosophy of Science, Vol. 1. Minneapolis: University of Minnesota Press.

Shapin, S. & Schaffer, S. (1985). *Leviathan and the Air-Pump: Hobbes, Boyle and the Experimental Life*. Princeton, NJ: Princeton University Press.

Smart, J. J. C. (1963). *Philosophy and Scientific Realism*. London: Routledge & Kegan Paul.

Sprevak, M. (2009). Extended Cognition and Functionalism. *The Journal of Philosophy*, 106, 503–27.

Sprevak, M. (2013). Fictionalism about Neural Representations. *The Monist*, 96 (4), 539–60.

Stanley, J. (2001). Hermeneutic Fictionalism. *Midwest Studies in Philosophy* 25 (1), 36–71.

Suárez, M. (ed.) (2009). *Fictions in Science: Philosophical Essays on Modeling and Idealization*. Abingdon, Oxfordshire: Routledge.

Tanney, J. (2009). Rethinking Ryle: A Critical Discussion of the Concept of Mind. In G. Ryle (1949/2009), *The Concept of Mind: 60th Anniversary Edition* (pp. ix–lix). Abingdon, Oxfordshire: Routledge.

Toon, A. (2010). 'The Ontology of Theoretical Modelling: Models as Make-Believe'. *Synthese*, 172(2), 301–15.

Toon, A. (2011). 'Playing with Molecules'. *Studies in History and Philosophy of Science*, 42(4), 580–9.

Toon, A. (2012). *Models as Make-Believe: Imagination, Fiction and Scientific Representation*. Basingstoke: Palgrave Macmillan.

Toon, A. (2015). Where Is the Understanding? *Synthese*, 192 (12): 3859–75.

Toon, A. (2016). Fictionalism and the Folk. *The Monist*, 99: 280–95.

Toon, A. (2019). Review of *As If: Idealization and Ideals*, by Kwame Anthony Appiah. *Mind*, 129 (513): 275–283.

Vaihinger, H. (1924). *The Philosophy of 'As If': A System of the Theoretical, Practical, and Religious Fictions of Mankind*. Abingdon, Oxford: Routledge.

Van Fraassen, B. (1980). *The Scientific Image*. Oxford: Oxford University Press.

Wallace, M. (2016). Saving Mental Fictionalism from Cognitive Collapse. *Res Philosophica*, 93(2), 405–24.

Wallace, M. (2022). Mental Fictionalism. In T. Demeter, T. Parent, & A. Toon (eds), *Mental Fictionalism: Philosophical Explorations* (pp. 27–51). Abingdon, Oxfordshire: Routledge. (Revised from an unpublished manuscript, first circulated in 2007.)

Walton, K. L. (1990). *Mimesis as Make-Believe: On the Foundations of the Representational Arts*. Harvard: Harvard University Press.

REFERENCES 137

Walton, K. L. (1993). Metaphor and Prop Oriented Make-Believe. *European Journal of Philosophy*, 1(1), 39–56.

Walton, K. L. (2000). Existence as metaphor? In A. Everett and T. Hofweber (eds.), *Empty Names, Fiction, and the Puzzles of Non-Existence* (pp. 69–94). Stanford: CSLI Publications. Reprinted in Walton, K., *In Other Shoes: Music, Metaphor, Empathy, Existence* (pp. 89–113). New York: Oxford University Press. (Page numbers cited in text refer to reprinted version.)

Weiskopf, D. (2008). Patrolling the Mind's Boundaries. *Erkenntnis*, 68(2), 265–76.

Weiskopf, D. (2009). The Plurality of Concepts. *Synthese*, 169(1), 145.

Wheeler, M. (2005). *Reconstructing the Cognitive World*. Cambridge, MA: MIT Press.

Wheeler, M. (2011). In Search of Clarity about Parity. *Philosophical Studies*, 152, 417–25.

Wilkinson, S. (2020). The Agentive Role of Inner Speech in Self-Knowledge. *Teorema*, 39(2), 7–26.

Wilson, R. A. (2004). *Boundaries of the Mind: The Individual in the Fragile Sciences: Cognition*. Cambridge: Cambridge University Press.

Wittgenstein, L. (1953). *Philosophical Investigations*. Oxford: Blackwell. (Translated by G. E. M. Anscombe.)

Yablo, S. (1998). Does Ontology Rest on a Mistake? *Proceedings of the Aristotelian Society, Supplementary Volume*, 72, 229–61.

Yablo, S. (2000). A Paradox of Existence. In A. Everett and T. Hofweber (eds), *Empty Names, Fiction, and the Puzzles of Non-Existence* (pp. 275–312). Stanford: CSLI Publications.

Yablo, S. (2001). Go Figure: A Path through Fictionalism. *Midwest Studies in Philosophy*, 25(1), 72–102.

Yablo, S. (2005). The Myth of the Seven. In M. E. Kalderon (ed.), *Fictionalism in Metaphysics* (pp. 88–115). Oxford: Oxford University Press.

Ylikoski, P. (2009). The Illusion of Depth of Understanding in Science. In H. de Regt, S. Leonelli, & K. Eigner (eds), *Scientific Understanding: Philosophical Perspectives* (pp. 100–19). Pittsburgh: Pittsburgh University Press.

Index

For the benefit of digital users, indexed terms that span two pages (e.g., 52–53) may, on occasion, appear on only one of those pages.

4E cognition (*see also* extended mind thesis) 6–7, 85

aboutness (*see* intentionality)
abstracta vs. illata 43
accommodation vs. prediction 103–4
animal beliefs 42, 65, 99
artificial neural networks (*see* connectionism)
assertion 72–3
atoms, as real fictions 78–9, 82

Baker, L. R. 70, 72
behaviourism 33–6, 38–9, 50–1, 106–7
beliefs
    as extended 85–7, 90–1, 97–9
    fictionalist approach to 21–3, 28–30, 45–6, 88–9
    and folk psychology 9–11
    indeterminacy of 29–30, 97–9
    and knowledge 115–16
    occurrent vs. standing 20
Bennett, M. 45, 80
birdwatching 110–14
'blockbuster' 40–1
Blockhead 39–41
Block, N. 39–41
Bourne, C. 20n.3, 45, 71n.1
Brandom, R. 71–3

Caddick Bourne, E. 20n.3, 45, 71n.1
Carruthers, P. 55, 58–9
Cartesianism 2, 11–15, 26–7, 36, 38, 49–52, 123
category mistakes 2, 26–8, 36–8
causation, mental 10, 13–14, 48–52
Chalmers, D. 85–7
Churchland, P. M. 14–15, 70, 103–4, 106–7
Clark, A. 85–7, 95
cognitive bloat 96–9
cognitive collapse 68–75
cognitive science 3–4, 6–7, 12–15, 30–1, 82, 95–6
computationalism (*see* computational theory of mind)

computational theory of mind 12–15, 49–50
concepts
    as extended 111–13, 124–5
    fictionalist approach to 109–15, 124–6
    pluralism about 114
    possession of 109
    structure of 113–14
conceptual structure (*see* concepts)
connectionism 13–15, 106–7
consciousness (*see* self-knowledge)
content of thought (*see* intentionality)
copy principle 125–6
Crane, T. 59n.1, 87

dead metaphors (*see* metaphor)
decisions 21, 57–9
Demeter, T. 6
Dennett, D. 28–30, 39–46, 70–1, 95–6
derived intentionality (*see* intentionality)
Descartes, R. (*see also* Cartesianism) 2
desires 9–11, 22, 47–8, 50, 90
Draaisma, D. 3, 21–2
Drayson, Z. 31n.4
dualism 2, 12, 28, 49

Elgin, C. 119
eliminativism 14–15, 31–2, 70, 95, 106–7
eliminative materialism (*see* eliminativism)
empiricism 125–6
epistemology 105–22
explanation
    and fictions 76–9, 82–3
    folk psychological 48–52
    and inference (*see* inference to the best explanation)
    and success (*see* no miracles argument)
extended beliefs (*see* beliefs)
extended concepts (*see* concepts)
extended knowledge (*see* knowledge)
extended understanding (*see* understanding)
extended mind thesis
    and belief 85–7, 90–1, 97–9
    and cognitive bloat 96–9

140 INDEX

extended mind thesis (*cont.*)
  and common sense 92–3
  and concepts 111–13, 124–5
  and intentionality 93–5
  and knowledge 117–19, 124–5
  fictionalist interpretation of 89–99
  and non-representational tools 94–5,
    117–18
  and representationalism 87
  and understanding 121–2, 124–5
external representations (*see* representations)

fairy tales 80–1
Farkas, K. 59n.1, 117–18
fictions
  real fictions 78–83
  semi-fictions 78–9
  vs. hypotheses 76–7
fictionalism
  hermeneutic vs. revolutionary 108–9
  mathematical 5, 46–8
  meta-fictionalism 46–7
  prefix vs. pretence-fictionalism 46–8
Fodor, J. 12–14, 51, 62, 83, 100–2
folk psychology 9–11
  and explanation 48–52
  fictionalist approach to 18–30, 51–2, 87–93
  success of 14–15, 99–104
  as a theory of inner machinery 11–15

games of make-believe 15–18, 55, 71–5, 80–1,
    102–3
ghost in the machine 2, 128
glue and trust conditions 86, 88, 112, 116
Griceanism 61, 65–6

Hacker, P. 45, 80
Hacking, I. 123
Hampshire, S. 2–3
Haugeland, J. 49, 67–8
hermeneutic fictionalism (*see* fictionalism)
Hume. D. 122–6
Hutto, D. 31n.4, 100
hypotheses (*see* fictions)

idealism 12, 38
illata (*see* abstracta vs. illata)
imagination 69–72
indeterminacy of beliefs (*see* beliefs)
inference to the best explanation 77–9, 82–3,
    99–100
inner representations (*see* mental
    representations)
inner speech 19n.2, 56–9
instrumentalism 38–46, 50–1

intentionality 10, 60–1
  and the extended mind thesis 93–5
  fictionalist approach to 63–8, 71–5, 93–5
  of language 65–8, 93–5
  original vs. derived 60–1, 93–5
  of thought 10, 63–5, 68, 93–5
intentional stance (*see also* Dennett, D.) 41–6,
    51, 70–1
intentional systems theory 38–9, 42
introspection (*see* self-knowledge)

Jones, myth of 18–21, 63–4, 67–8, 75, 88
judgement 20–1, 26–7, 57–9

knowledge
  and belief 115–18
  as extended 117–19, 124–5
  fictionalist approach to 115–19, 124–5

Langland-Hassan, P. 73–5
language
  intentionality of 65–8, 93–5
  and mental states 64–5, 84–5, 123
language of thought (*see also* Fodor, J.) 13
Law, J. 110, 112–13
logical geography 36–7
Lynch, M. 110, 112–13

make-believe
  content-orientated vs. prop-orientated 16–18,
    23–4
  games of 15–18, 48, 55, 71–5, 80–1, 102–3
mathematical fictionalism (*see* fictionalism)
material culture (*see* representations *and* tools)
materialism 1–2, 12–13, 38
meaning (*see* intentionality)
Melser, D. 3
memory
  as extended 85–7, 90–1, 97–9, 101–2
  fictionalist approach to 21–3, 65, 88–9,
    97–9, 102
mental causation (*see* causation, mental)
mental representation (*see* representations)
meta-fictionalism (*see* fictionalism)
metaphor
  aptness of 30, 65, 89–90, 97–9
  pretence analysis of 15–18
  primary and secondary domain 18–20, 32
  as representationally essential 34–5, 100
missing shade of blue 125–6
models, scientific 5–6, 76–9, 82, 100–1
myth of Jones (*see* Jones, myth of)

negative transfer 96–8
neural network modelling (*see* connectionism)

INDEX   141

no miracles argument 77–9, 99–104
norms (*see also* principles of generation) 67–8,
     71–5

objective knowledge 107–9, 115–16, 123
occurrent belief (*see* judgement)
occurrent vs. standing states 20
original intentionality (*see* intentionality)
Otto and Inga 85–90, 94, 96–7, 101, 118

pain 53
parity principle 86, 91
phenomenological objection 24–30
paradox of mechanical reason 49–50
Parent, T. 46
Popper, K. 107, 123
prefix-fictionalism (*see* fictionalism)
pretence 15–18, 71–5, 80–1
primary domain (*see* metaphor)
principles of generation 15–16, 23, 48
propositional attitudes 3–4, 9–10, 20, 57, 116–17
prop-oriented make-believe (*see* make-believe)

realism
     about mental representations 11–14, 61–3,
          75–83, 99–104
     about mental states 30–2
     scientific 77–8, 82–3, 99–100, 103–4
real patterns 31
reductionism 34–6, 38, 100
regulative aspect of folk psychology 102–3
representations
     mental 11–14, 61–3, 75–83, 87
     public 4, 19–22, 65–8, 72–5, 84–5, 87–91,
          108–22
representational theory of mind 3–4, 11–14,
          49–50, 61–3, 77, 81–2
     and the extended mind thesis 87
representationalism (*see* representational theory
          of mind)

revolutionary fictionalism (*see* fictionalism)
Ross, A. 20n.3
rules (*see* norms *and* principles of generation)
Ryleans (*see* Jones, myth of)
Ryle, G. 2–3, 26–8, 36–8, 51, 53–5, 72, 107–8,
          115–17, 123–4

Schwitzgebel, E. 36
scientific models (*see* models, scientific)
scientific realism (*see* realism)
secondary domain (*see* metaphor)
Searle, J. 123–4
self-knowledge 20, 52–9
Sellars. W. 18–20, 63–4, 67, 72–3
self-refutation (*see* cognitive collapse)
sensations 4–5, 53–6
silent soliloquy (*see* inner speech)
silly questions 28–9, 92–3
social norms (*see* norms)
Sprevak, M. 48, 82–3
standing states (*see* occurrent vs.
          standing states)
syntax vs. semantics 12–13, 49–50

third world (*see* objective knowledge)
thought 3–4, 9–10, 57, 59n.1
tools 67, 89–91, 122–6

understanding
     as extended 121–2, 124–5
     fictionalist approach to 119–22, 124–5
     and knowledge 119

Vaihinger, H. 76–80, 82

Wallace, M. 6, 46, 48
Walton, K. 5–6, 15–18, 71–2
Wittgenstein, L. 67–8, 73–4, 80–1

Yablo, S. 28–9, 34–5, 51